DNA Decipher Journal

Volume 11 Issue 1

October 2021

Geometric Representation & Physics of the Genetic Code

Editors:

Huping Hu, Ph.D., J.D.

Maoxin Wu, M.D., Ph.D.

Advisory Board

Matti Pitkanen, Ph.D., Independent Researcher, Finland
Andrei Khrennikov, Professor, In'tl Center for for Mathematical Modeling, Linnaeus Univ., Sweden
Massimo Pregnolato, Professor, Quantumbiolab, Dept. of Drug Sciences, Univ. of Pavia, Italy
Chris King, Independent Researcher, New Zealand
Graham P. Smetham, Independent Researcher, United Kingdom

ISSN: 2159-046X　　　DNA Decipher Journal　　　www.dnadecipher.com
Published by QuantumDream, Inc.

Table of Contents

Articles

Algebraic and Geometric Representations of the Genetic Code *Richard L Amoroso, Peter Rowlands*	01-10
Is Genetic Code Part of Fundamental Physics in TGD Framework? *Matti Pitkanen*	11-20
Homeostasis as Self-organized Quantum Criticality *Matti Pitkanen & R. Rastmanesh*	21-38
TGD View about Language *Matti Pitkanen & R. Rastmanesh*	39-70

Explorations

The TGD Based View about Dark Matter at the Level of Molecular Biology *Matti Pitkanen & R. Rastmanesh*	71-92
MeshCODE Theory from TGD Point of View *Matti Pitkanen*	93-103

In Memoriams

Peter P. Gariaev (1942 - 2020): Discoverer of Phantom DNA Effect & Founder of "Wave Genetics" *Huping Hu & Maoxin Wu*	104-106
Iona Miller (1949 - 2021): Multitalented Writer, Artist & Visionary *Huping Hu & Maoxin Wu*	107-108

Article

Algebraic and Geometric Representations of the Genetic Code

Richard L Amoroso[*,1], Peter Rowlands[2]

[1]Noetic Advanced Studies Institute, USA
[2]University of Liverpool, UK

Abstract

Algebraic and geometric representations of the genetic code are used to show their functions for coding amino acids. The algebra is a 64-part vector quaternion combination, and the related geometry is based on the structure of the regular icosidodecahedron. An almost perfect pattern suggesting that this is a biologically significant way of representing the genetic code that may lead to a deeper understanding of a relationship between geometry and teleological life principles of complex self-organization.

Keywords: Genetic code, triplet codons, amino acids, vector-quaternion algebra, teleology, Icosidodecahedron.

1. Introduction

We explore an interesting way to represent the genetic code using a correspondence between algebra and geometry. The algebraic component is based on an Icosian calculus with a non-commutative algebraic structure discovered by William Rowan Hamilton in 1856, which he called quaternions. In modern terms, Hamilton produced a group presentation of the icosahedral rotation group by generators and relations.

Hamilton's discovery was derived from his attempts to find an algebra of 'triplets' that he believed would reflect the three Cartesian axes in a manner extending the complex numbers, which took the form, $i^2 = j^2 = k^2 = ijk = -1$. The symbols of the icosian calculus can be equated to moves between vertices on a dodecahedron.

2. The Algebraic Representation

In previous work [1-3] we have used various mathematical structures to represent the genetic code, including a 64-part vector quaternion algebra, which is isomorphic to the algebra of the quantum mechanical Dirac equation, and a combination of the faces and vertices of a regular icosidodecahedron. Here, we aim to show that it is possible to represent the codon structures both algebraically and geometrically in a way that relates to their function in coding for amino acids.

---

[*]Correspondence: Prof. Richard L. Amoroso, Director of Physics Lab., Noetic Advanced Studies Institute, Utah, USA. https://orcid.org/0000-0003-2405-9034; http://www.noeticadvancedstudies.us E-mail: amoroso@noeticadvancedstudies.us

It is based on a vector-quaternion algebra whose units can be represented as follows [4-6]:

i	j	k	vector
i			pseudoscalar
i	j	k	quaternion
1			scalar

They can be considered as the units of two spaces: ('real', constructed from i, j, k) and ('vacuum', constructed from i, i, j, k, 1). In principle, any self-organizing system, whether physical, chemical or biological, forms a space, which has a kind of distorted mirror image in another 'space' representing the rest of the universe. The double space creates the entire combination of system and 'vacuum' as a zero totality.

In principle, any self-organizing system, whether physical, chemical or biological, forms a space, which has a kind of distorted mirror image in another 'space' representing the rest of the universe. The double space creates the entire combination of system and 'vacuum' as a zero totality. The algebraic structure has an exact parallel with a geometric one which can be represented using Platonic or Archimedean solids in which each structure has a dual which could be imagined as constructed in another space.

The units of the vector-quaternion algebra constructing the double space can be represented as follows:

i	j	k	i*i	i*j	i*k	i	1	–i	–j	–k	–i*i	–i*j	–i*k	–i	–1
i	j	k	i*i	i*i	i*k			–i	–j	–k	–i*i	–i*i	–i*k		
i*i	i*j	i*k	i*i*i	i*i*i	i*i*k			–i*i	–i*j	–i*k	–i*i*i	–i*i*i	–i*i*k		
j*i	i*j	j*k	i*j*i	i*j*i	i*j*k			–j*i	–i*j	–j*k	–i*j*i	–i*j*i	–i*j*k		
k*i	k*j	k*k	i*k*i	i*k*i	i*k*k			–k*i	–k*j	–k*k	–i*k*i	–i*k*i	–i*k*k		

An alternative ordering would separate the four complex numbers from 12 nilpotent structures, each formed from 5 units. Here, we create a subset of 60 units, which has significance in the dodecahedral and icosahedral representations and in Hamilton's Icosian calculus:

1					–1				
ii	ij	ik	ik	j	–ii	–ij	–ik	–ik	–j
ji	jj	jk	ii	k	–ji	–jj	–jk	–ii	–k
ki	kj	kk	ij	i	–ki	–kj	–kk	–ij	–i

i					–i				
iii	iij	iik	ik	j	–iii	–iij	–iik	–ik	–j
iji	ijj	ijk	ii	k	–iji	–ijj	–ijk	–ii	–k
iki	ikj	ikk	ij	i	–iki	–ikj	–ikk	–ij	–i

One way of generating the 64 units is by taking the product of 4 options × 4 options × 4 options, as is done in the case of the genetic code, where each of three bases may be U (or T), G, A or C. To represent this algebraically, we may use the vector units i, j, k and 1 for the options U, G, A and C for the first base. Then, for the second base, we may represent U, G, A and C by the quaternion units i, j, k and 1. Then U, G, A and C on the third base may be represented by the units of complex algebra 1, i, −1, −i. Using three different algebras (vectors, quaternions and complex numbers) allows us to track the three bases individually according to their positions in the codon.

We have previously grouped the amino acids produced by the genetic code mechanism according to the second base in the codon which produced it. The second base seems, in this respect, the most important, and the third base the least, becoming in some sense almost redundant. Using this division, the 64 codons fall naturally into 4 groups of 16:

amino acid	codon	first base	second base	third base

Group I

amino acid	codon	first base	second base	third base
Phe	UUU	i	i	1
	UUC	i	i	i
Leu	UUA	i	i	-1
	UUG*	i	i	$-i$
	CUU	1	i	1
	CUC	1	i	i
	CUA	1	i	-1
	CUG*	1	i	$-i$
Val	GUU	j	i	1
	GUC	j	i	i
	GUA	j	i	-1
	GUG	j	i	$-i$
Ile	AUU	k	i	1
	AUC	k	i	i
	AUA*	k	i	-1
Met	AUG*	k	i	$-i$

Group II

amino acid	codon	first base	second base	third base
Cys	UGU	i	j	1
	UGC	i	j	i
Trp	UGG	i	j	$-i$
STOP	UGA	i	j	-1
Gly	GGU	j	j	1
	GGC	j	j	i
	GGA	j	j	-1
	GGG	j	j	$-i$
Ser	AGU	k	j	1
	AGC	k	j	i
Arg	CGU	1	j	1
	CGC	1	j	i
	CGA	1	j	-1
	CGG	1	j	$-i$
	AGA	k	j	-1
	AGG	k	j	$-i$

Group III

amino acid	codon	first base	second base	third base
STOP	UAA	i	k	-1
	UAG	i	k	$-i$
Tyr	UAU	i	k	1
	UAC	i	k	i
Asp	GAU	j	k	1
	GAC	j	k	i
Glu	GAA	j	k	-1
	GAG	j	k	$-i$
Lys	AAA	k	k	-1
	AAG	k	k	$-i$
Asn	AAU	k	k	1
	AAC	k	k	$-i$
His	CAU	1	k	1
	CAC	1	k	i
Gln	CAA	1	k	-1
	CAG	1	k	$-i$

Group IV

amino acid	codon	first base	second base	third base
Ser	UCU	i	1	1
	UCC	i	1	i
	UCA	i	1	-1
	UCG	i	1	$-i$
Ala	GCU	j	1	1
	GCC	j	1	i
	GCA	j	1	-1
	GCG	j	1	$-i$
Thr	ACU	k	1	1
	ACC	k	1	i
	ACA	k	1	-1
	ACG	k	1	$-i$
Pro	CCU	1	1	1
	CCC	1	1	i
	CCA	1	1	-1
	CCG	1	1	$-i$

The

asterisks represent codons that can act as a START. Notably they are all in the same group. Conveniently also (though this is mainly an artefact of our representation) all the START and STOP codons are represented by negative units. The way that the various structures are relevant to the formation of amino acids will become clearer in the following table:

Group I
Phenylalanine
  UUU   UUC
   ii iii
Leucine
  UUA   UUG*   CUU   CUC   CUA   CUG*
  $-ii$ $-iii$ i ii $-i$ $-ii$
Valine
  GUU   GUC   GUA   GUG
   ij iij $-ij$ $-iij$
Isoleucine
  AUU   AUC   AUA*
   ik iik $-ik$
Methionine
  AUG*
  $-iik$

Group II
Cysteine
  UGU UGC
   ji iji
Tryptophan
  UGG
  $-iji$
STOP
  UGA
  $-ji$
Glycine
  GGU   GGC   GGA   GGG
  $-jj$ $-ijj$ jj ijj
Serine
  AGU   AGC
   jk ijk
Arginine
  CGU   CGC   CGA   CGG   AGA   AGG
   j ij $-j$ $-ij$ $-jk$ $-ijk$

Group III
STOP
  UAA   UAG
  −*k*i   −i*k*i
Tyrosine
  UAU   UAC
  *k*i   i*k*i
Aspartate
  GAU   GAC
  *k*j   i*k*j
Glutamate
  GAA   GAU
  −*k*j   −i*k*j
Lysine
  AAA   AAG
  −*k*k   −i*k*k
Asparagine
  AAU   AAC
  *k*k   i*k*k
Histidine
  CAU   CAC
  k   i*k*
Glutamine
  CAA   CAG
  −*k*   −i*k*

Group IV
Serine
  UCU   UCC   UCA   UCG
  i   ii   −i   ii
Alanine
  GCU   GCC   GCA   GCG
  j   ij   −j   −ij
Threonine
  ACU   ACC   ACA   ACG
  k   i*k*   −*k*   −i*k*
Proline
  CCU   CCC   CCA   CCG
  1   i   −1   −i

We can see the pattern is nearly perfect; illustrated, particularly by the complete regularity of Groups II and IV. Only the two serine codons in Group II are anomalous (not in their biological group), and they will be in any arrangement. Arginine and serine in this group seem to each have two codons that could, originally, have coded a different amino acid. Almost certainly, the codons for arginine and serine have become mixed at some stage in biological evolution. (We may note also that arginine seems to be an exception to the general tendency for the more complicated amino acid molecules to be coded using fewer codon pathways.) The table we have given is only for one species, and it may be that evolutionary drift may be determined to some extent by the variations in the patterns from the assumed perfect norm. The codon for tryptophan, notably, can become the STOP codon in some species, and vice versa.

3. The Geometrical Representation

Algebraic and geometrical structures are fundamentally dual. Where there is an algebra, there is also a geometry, and vice versa. It is easy to show that this is the case here. The four groups of codons can now be represented on the faces of a regular icosidodecahedron, divided into four equal sections. (We could use the combined faces plus vertices of a dodecahedron or icosahedron.) The negative units are not shown in the figures, but can be assumed either to be represented on the corresponding vertices of the dual rhombic triacontahedron, or on the inner, rather than outer, surface of the icosidodecahedron.

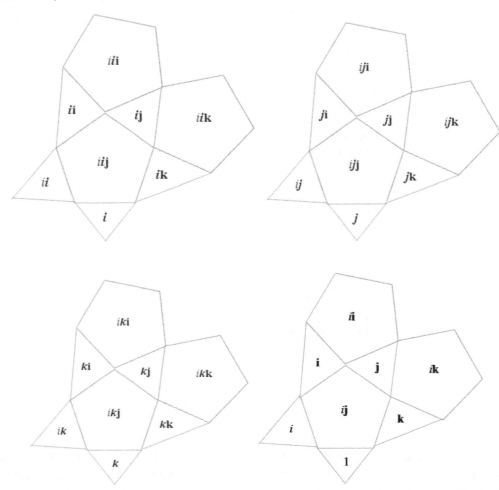

Figure 1. Algebraic geometry for representing codons on the icosidodecahedron.

The codons can be represented on these diagrams in the form:

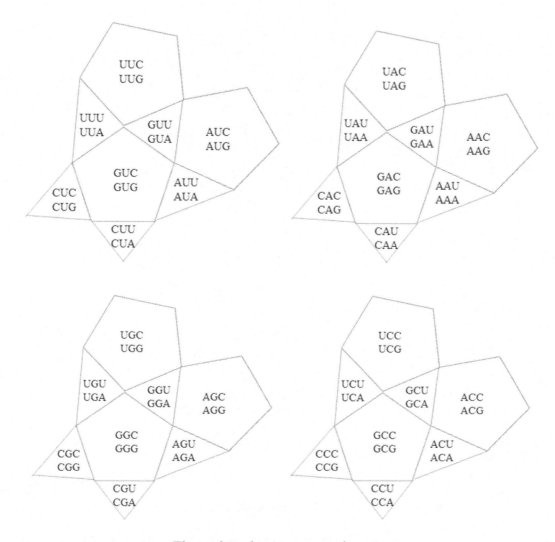

Figure 2. codons representations

The amino acids coded can be represented as follows:

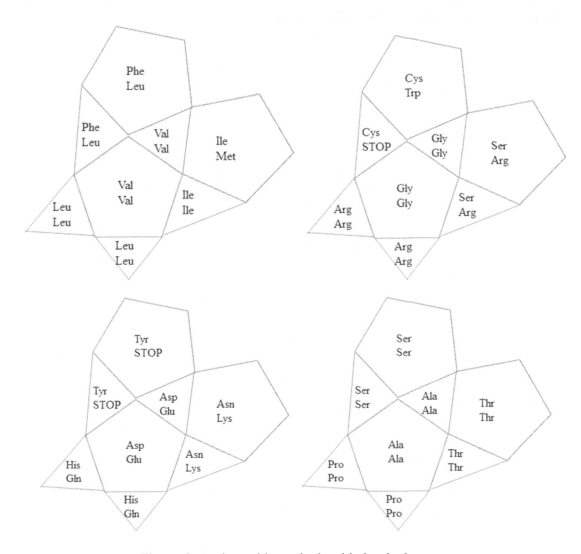

Figure 3. Amino acids on the icosidodecahedron.

Here, the position of serine in two separate groups gives an idea of how the four sections might be connected. One of the significant aspects of the first three sections in alternative arrangement (below) is that the algebraic units of the three pentagons in each, combined with those of the two outer triangles on the lowest pentagon form the basis of a nilpotent structure, such as we find in the amplitude term in the nilpotent Dirac equation ($ikE + iip_x + ijp_y + ikp_z + jm$). At the same time, the five triangles taken together form the basis of another nilpotent structure. So, the three inner triangles and the three pentagons display a duality in that either group can be used with the two outer triangles to generate a set of nilpotent units (though with their roles switched in the two cases). This provides another way of generating 12 nilpotent structures from the algebra. Even using the first version of the icosidodecahedral sections, we can connect these triangles making up the nilpotent units with the upper pentagons, and so maintain the nilpotent structure. Nilpotency is one of the assumed bases of the overall pattern that we have described as Nature's code, and it appears to be the means by which a self-organizing system connects with its external environment. Its presence in these geometric structures indicates the real importance of geometry in the genetic code as a route towards self-organization.

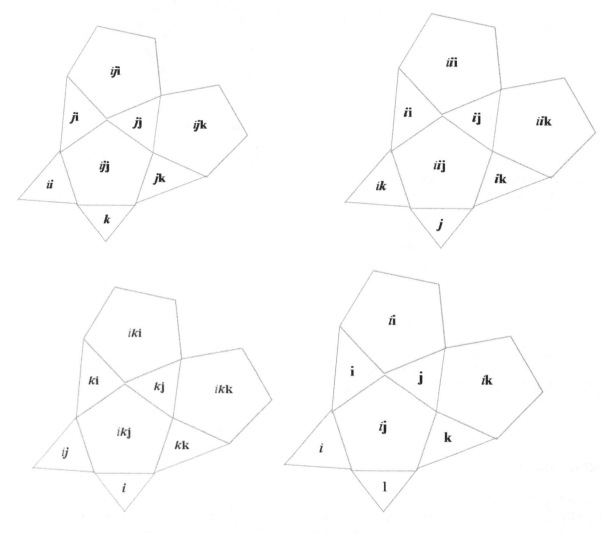

Figure 4. Quaternion algebraic representation.

Appendix: Note on the Regular Icosidodecahedron

Geometrically, an icosidodecahedron is a polyhedron with twenty (icosi) triangular faces and twelve (dodeca) pentagonal faces. An icosidodecahedron has 30 identical vertices, with two triangles and two pentagons meeting at each vertex. It also has 60 identical edges, each separating a triangle from a pentagon. Because of this, it is one of the Archimedean solids

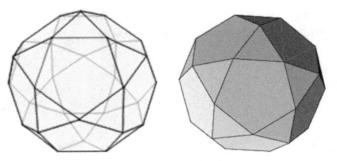

Figure 5. Two views of the icosidodecahedron.

All Archimedean solids can be produced from Platonic solids, by 'cutting the edges' of the platonic solid. Likewise, Platonic solids can be turned into Archimedean solids by following a series of rules for their construction.

Interestingly, in Cartesian coordinates, the vertices of an icosidodecahedron with unit edges are given by the even permutations of

$$(0, 0 \pm \varphi)$$
$$\left(\pm\frac{1}{2}, \pm\frac{\varphi}{2}, \frac{\varphi^2}{2}\right),$$

where φ is the golden ratio, $\frac{1+\sqrt{5}}{2}$ [7].

Received June 8, 2021; Accepted June 26, 2021

References

[1] Hill, V.J. and Rowlands, P. (2008) Nature's code, AIP Conference Proceedings, 1051, 117-126.
[2] Hill, V.J. and Rowlands, P. (2010) Nature's fundamental symmetry breaking, International Journal of Computing Anticipatory Systems, 25, 144-159.
[3] Hill, V.J. and Rowlands, P. (2010) The numbers of nature's code, International Journal of Computing Anticipatory Systems, 25, 160-175.
[4] Rowlands, P. (2007) Zero to Infinity: The Foundations of Physics, Singapore and Hackensack, N.J., World Scientific.
[5] Rowlands, P. (2010) Dual vector spaces and physical singularities, AIP Conference Proceedings, 1316, 102-111.
[6] Amoroso, R.L., Rowlands P., Kauffman, L.H. (2013) Exploring novel cyclic extensions of Hamilton's dual-quaternion algebra, in R.L. Amoroso, L.H. Kauffman, P. Rowlands (eds.) The Physics of Reality Space, Time, Matter, Cosmos, Proceedings of the 8th Symposium Honoring Mathematical Physicist Jean-Pierre Vigier, pp. 81-92, Singapore: World Scientific; https://vixra.org/pdf/1711.0468v1.pdf.
[7] Weisstein, E. W. (2010) Icosahedral Group, MathWorld, A Wolfram Web Resource; https://mathworld.wolfram.com/IcosahedralGroup.html.

Is Genetic Code Part of Fundamental Physics in TGD Framework?

Matti Pitkänen [1]

Abstract

Topological Geometrodynamics (TGD) leads to 3 basic realizations of the genetic code. Besides the chemical realization, there is a realization in terms of dark proton sequences (dark nuclei) with 3-proton state representing codon. Ordinary DNA strands would be accompanied by dark magnetic flux tubes carrying the dark proton triplets. Also RNA, amino-acids and tRNA would have dark proton analogs. The third realization is in terms of dark photon triplets and involves the notion of bio-harmony modelled in terms of icosahedral and tetrahedral geometries with 3-chords of light assigned to the triangular faces of icosahedron and tetrahedron. 12-note scale is realized as a Hamiltonian cycle for icosahedron with the step between nearest neighbor vertices for the cycle realised as a quint scaling of frequency. The 3-chords correspond to the triangular faces of the icosahedron. Also tetrahedral realization of 4-note scale is necessary in order to obtain genetic code. DNA codons correspond to triangular faces and the orbit of a given triangle under the symmetries of the icosahedral harmony correspond to DNA codons coding for the amino acid assigned with the orbit. Vertebrate genetic code emerges as a prediction. Codon corresponds to 6 bits: this is information in the usual computational sense. Bio-harmony codes for a mood: emotional information related to emotional intelligence. Bio-harmony would be a fundamental representation of emotional information realized already at the molecular level. Dark photon 3-chords and more generally, 3N-chords representing genes, would mediate interaction between various realizations.

Why both icosahedron and tetrahedron? Could this relate to the fact that the tesselations of a hyperbolic space H^3 are in fundamental role in quantum TGD. There is one particular tesselation known as tedtrahedral-icosahedral honeycomb. Could the genetic code be realized at the level of fundamental physics as a tedrahedral-icosahedral tesselation of H^3 emerging as a cognitive representation for space-time surfaces at the level of M^8 and by $M^8 - H$ duality also in $H = M^4 \times CP_2$. Biological realization could be only one particular realization. It should be possible to unify all models of the genetic code to single model so that the codon as a dark proton triplet is assigned to a representation as an "active" triangle of icosahedron or tetrahedron containing at it vertices dark protons defining the same codon as the triangle as 3-chord for a given icosahedral harmony. Could these "activated" triangles be assigned with tetrahedral-icosahedral tesselation? Could genes correspond to sequences of these icosahedron-tetrahedron pairs at magnetic flux tubes defined dark generatic code? Why there should be 3 icosahedral harmonies and one tetrahedral harmony? There is a partial answer to this question. The correspondence with 64 dark proton triplets representing codons and triangles requires 3 icosahedral harmonies. What distinguishes stop codons from other codons? It turns out that stop codons could be dark proton triplets for which the corresponding triangle does not exist in tetrahedral-icosahedral realization! The lack of a dark proton triplet would mark the end of gene.

1 Introduction

Topological Geometrodynamics (TGD) proposes 3 basic realizations of the genetic code [16]. The first realization is the standard chemical realization. The second realization is in terms of dark proton sequences (dark nuclei) with proton triplet representing a codon. Ordinary DNA strands would be accompanied by dark magnetic flux tubes carrying the dark proton triplets. Also RNA, amino-acids and tRNA would have dark proton analogs.

The third realization is in terms of dark photon triplets and involves the notion of bio-harmony described in terms of icosahedral and tetrahedral geometries with 3-chords of light (perhaps also sound)

---

[1] Correspondence: Matti Pitkänen http://tgdtheory.com/. Address: Rinnekatu 2-4 A8, 03620, Karkkila, Finland. Email: matpitka6@gmail.com.

assigned to the triangular faces of icosahedron and tetrahedron. 12-note scale is realized as a Hamiltonian cycle for icosahedron with the step between nearest neighbor vertices for the cycle realised as quin (scaling of frequency by factor 3/2). The 3-chords correspond to the triangular faces of the icosahedron. Also tetrahedral realization of 4-note scale is necessary in order to obtain genetic code. DNA codons correspond to triangular faces and the orbit of a given triangle under the symmetries of the bio-harmony corresponds to DNA codons coding for the amino acid assigned with the orbit. Vertebrate genetic code emerges as a prediction.

The 3-chords of dark photon triangles are assigned with the triangular faces of icosahedron and tetrahedron [13, 16, 22] such that their corners are labelled by the notes of the 12- and 4-note scales realized as a icosahedral and tetrahedral Hamiltonian cycles, which are closed paths connecting vertex to neighboring vertex and going through every vertex once.

Genetic code corresponds to a fusion of tetrahedral harmony with 4 chords and of 3 icosahedral harmonies with 20 3-chords having as group of symmetries Z_6, Z_4 and Z_2 - Z_2 can correspond either to reflection or rotation by π. There are also 6 disharmonies without any symmetries (Z_1) with single DNA codon coding for single amino-acid. There is a considerable number of different icosahedral harmonies and the 3 icosahedral harmonies can be in different key so that a large number of bio-harmonies is possible [22]. The details of the model of bio-harmony are not completely fixed. In particular, the understanding of stop codons is not completely satisfactory. The small deviations from the vertebrate code (say bacteria and mitochondria) could be understood as being due to the incomplete mimicry of the dark code by chemical code in accordance with the idea that the mimicry has gradually evolved more complete.

Dark photon 3-chords mediate interaction between various realizations. Both dark proton and dark photon triplets would be dynamical units analogous to protons as color confined states of 3 quarks and in the adelic vision the notion of color confinement is replaced with Galois confinement [22, 24]. Also genes could be seen as Galois confined states of 3N dark protons and dark photons. 3N-photon exchange would be realized as 3N-fold frequency - and energy resonance (mere energy resonance) between dark levels with the same value (different values) of h_{eff}. The possibility to modify the value of h_{eff} for flux tube makes it possible to have for a given codon single resonance energy [24, 25, 26].

There are several questions relating to the bio-harmony.

1. The gluing of icosahedron and tetrahedron along the face looks ugly in the original model. Why both icosahedron and tetrahedron and why the gluing? The recent progress with $M^8 - H$ duality [19, 20] suggests an answer. The tesselations (honeycombs) of hyperbolic 3-space H^3 appear at the fundamental level and induce sub-tesselations of the magnetic flux tubes. One of these honeycombs- tetrahedral-icosahedral honeycomb (TIH)- involves all Platonic solids with triangular faces - tetrahedron, octahedron, and icosahedron. Could genetic code relate to TIH?

   Cognitive representation [15, ?, 18] as a set of points of space-time surface in the space of complexified octonions O_c with points having O_c coordinates in extension of rationals associated with the polynomial defining the space-time surfaces are central for for both quantum TGD and TGD inspired theory of cognition leading to adelic physics [14]. The cognitive representation is mapped to $H = M^4 \times CP_2$ by $M^8 - H$ duality [19, 20].

   Could the genetic code be realized at the level of fundamental physics as a TIH in H^3 emerging as a cognitive representation [15, 17, 18, 23] for the space-time surfaces in M^8 and by $M^8 - H$ duality also in $H = M^4 \times CP_2$. If so, the biological realization could be only one particular realization of the code.

2. Why there should ber 3 icosahedral harmonies and one tetrahedral harmony? There is a partial answer to this question. The correspondence with 64 dark proton triplets representing codons and triangles requires 3 icosahedral harmonies. What distinguishes stop codons from other codons? It turns out that stop codons could be dark proton triplets for which the corresponding triangle does not exist in THI realization! The lack of dark proton triplet would mark the end of the gene.

It should be possible to unify various TGD inspired models of genetic code to a single unified description. Is the time ripe for this?

1. The realizations in terms of dark protons and dark photons are related: dark photon 3N-plets would be emitted by dark proton 3N-plets in 3N-proton cyclotron transitions. In the 3N-resonance interaction with DNA, RNA, amino-acids, and tRNA the dark photon 3N-plet would transform to ordinary photons (bio-photons). Energy resonance could select the basic information molecules.

2. How the dark level interacts with the ordinary matter? Music expresses and creates emotions. Light 3-chords for a given bio-harmony could therefore represent an emotional state of MB (emotions as sensory perceptions of MB?). Fourier transform in terms of frequencies represents non-local holistic information and emotional information indeed is holistic information. Codons as units of 6 bits would represent ordinary temporily local, reductionistic information.

   Each emotional state corresponds to a particular collection of 3-chords as allowed chords of the bio-harmony and therefore the resonance occurs with different biomolecules or induces different transitions of these bio-molecules. Could this serve as a universal mechanism of bio-control? Could epigenesis as a control of DNA transcription rely on this mechanism? As a matter of fact, the model described in this article emerged from an attempt to understand epigenesis in the TGD framework.

3. Is it possible to unify all models of the genetic code to single model so that the representation of a codon as dark proton triplet is assigned to a representation as an "activated" triangle of icosahedron or tetrahedron of TIH containing at it vertices dark protons defining the same codon as the triangle as 3-chord for a given icosahedral harmony. Could these "activated" triangles be selected faces of TIH. Could genes correspond to sequences of these icosahedron-tetrahedron pairs at magnetic flux tubes?

In the sequel the questions raised above are discussed.

2 Genetic code and hyperbolic tesselations

Why 3 different icosahedral harmonies with symmetries Z_6, Z_4, and Z_2 plus one (there is only one) tetrahedral harmony is needed to get $3 \times 20 = 60 + 4$ chords in correspondences with 64 codons of the genetic code?

2.1 Hyperbolic tesselations and genetic code?

What comes into mind, are fundamental lattice like structures - tesselations - having as basic building bricks icosahedron and tetrahedron - at least these. This would make sensical to speak about gluing of tetrahedron to icosahedron, which looks a strange operation in the original formulation of the model.

1. Platonic solids correspond to finite tesselations at 2-sphere or equivalently 3-D solid polyhedrons in 3-D space Euclidian space E^3. Maybe one could answer the question by increasing dimension and by studying 3-D polyhedrons of 4-D space defining tesselations of the hyperbolic space H^3.

   By $M^8 - H$ duality [19, 20], these tesselations appear at the fundamental level TGD as cognitive representations since the 3-D mass shells with the geometry of H^3 appear naturally in the solutions of dynamical equations as algebraic equations at the level of M^8 identifiable as real section of complexified octonions O_c. The dynamics reduces to the associativity of the normal space of the space-time surface determined as a root for the real part of an octonionic polynomial obtained as an algebraic continuation of a real polynomial. Real part is defined in quaternionic sense by decomposing octonion to two quaternions in the same manner as a complex number is decomposed to its real and imaginary parts.

The algebraization of the octonionic counterpart of Dirac equation forces its identification as the counterpart of momentum space version of the ordinary Dirac equations and the identification of M^8 as an analog of momentum space so that space-time surface is analog of Fermi ball.

2. The tesselations of H^3 are analogs of lattices in an Euclidian momentum space E^3. In adelic physics they define cognitive representations providing unique discretizations of space-time surface both at the level of M^8 and H. $M^8 - H$ duality maps tesselations to their analogs of $H = M^4 \times CP_2$. Contrary to my long held belief, Uncertainty Principle forces the map to be instead of a direct identification an inversion for $M^4 \subset M^8 \to M^4 \subset H$ [19, 20]. Mass hyperboloids correspond in H to light-cone proper time constant sections of space-time surface: light-cone proper time defines Lorentz invariant cosmic time.

3. The tesselations of H^3 can have several different analogs of unit cells glued together along their 2-D faces. The positive curvature of sphere forces Platonic solids as tesselations of 2-sphere to be closed and be finite. H^3 as a negative curvature space does not allow a closure. This implies a large number of tesselations as infinite analogs of regular solid polyhedra. Both icosahedron, octahedron and tetrahedron have triangular faces so that they might allow gluing together for the simplest tesselations. Also more complex tesselations are possible.

2.1.1 Details about hyperbolic tessellations

Consider now in more detail some tessellations of H^3 possibly relevant for the bio-harmony [13, 16, 22] involving icosahedral and tetrahedral geometries.

Some basic concepts and notations are necessary to help the reader to understand the Wikipedia articles, which give detailed explanations and illustrations.

1. Regular polytopes are tesselations consisting of single polytope. There are subtle differences between tesselations and honeycombs: tesselations are more general than honeycombs. These differences are not relevant for what follows so that I will use both terms interchangeably.

2. Schläfli symbol [5] https://cutt.ly/7jagV1T $(p, q, r, ..)$ characterizes regular polytopes in both Euclidian spaces and hyperbolic spaces locally but does not tell anything about the object globally. For a 3-D regular polytope (p, q, r) in 4-D space (say tessellation of H^3 the faces have p vertices, q identical faces meet at given vertex, and r identical 3-cells meet along given edge. For instance, $(3, 5, 3)$ characterizes a regular tessellation having icosahedron as fundamental cells with 3 icosahedrons meeting along given edge.

3. Vertex figure [10] https://cutt.ly/yjagMQn) represents the neighboring vertices as seen from a given vertex. Formally it is defined by contracting all edges emanating from the vertex to their middle points and connecting these points by lines along faces. For a n-D polytope (p,r,s,..) the vertex figure is n-1-D polytope (r,s,...). For instance, for icosahedron (3,5) the vertex figure is (5) telling that 5 edges meet at vertex. For the regular honeycombs in H^3 the vertex figure is a regular polyhedron. For instance, for $(3, 5, 3)$ it is $(5,3)$ identifiable as dodecahedron. Second notation for the vertex figure is as the list of numbers of edges meeting at the vertices of the face: For icosahedron this list is 3.3.3.3.3 telling that the faces of the edge figure has 5 vertices at which edges meet.

4. Edge figure [10] (https://cutt.ly/djag9Q9) is the vertex figure of the vertex figure of the polytope. For D-dimensional polytope it is polytope of dimension D-2. For a regular polytope $(p, q, r, ..., s)$ the edge figure is $(r, ..., p)$: for Platonic solids (r,s) edge figure is () telling that two faces meet along a given edge. For the regular polytope (r,s,p) the edge figure tells the number of identical 3-cells meeting at given edge. For cubic lattice it is 4. For semiregular honeycombs the 3-cells need not be identical.

5. The notion of dihedral angle (see https://cutt.ly/vjs2OBI) is very useful in trying to understand whether a given tessellation of E^3 and H^3 is possible. Dihedral angle is defined as the angle between the faces of the polytope meeting along a given edge. For tetrahedron it is 120 °, for octahedron 90 ° and for icosahedron 138.19 °. Since at least 3 polyhedra must meet at a given edge, the sum of these angles must be smaller than 360 degrees in E^3. This prevents icosahedral tessellations in E^3.

   In H^3 negative curvature allows the sum to be larger than 360 ° (think of polygons at a saddle surface as a visualization) so that 3 icosahedra might meet at a given edge as indeed occurs for $(3,5,3)$ tessellation. The sum of the the dihedral angles of T, O, and I assignable to to tetrahedral-icosahedral honeycomb in H^3 is 348.19 ° and smaller than 360 ° but rather near to it.

6. An important notion is Coxeter group [2] (https://cutt.ly/FjdEJeG) acting as the symmetry group of the honeycomb. Coxeter group is generated by reflections meaning that honeycombs can be generated by reflections in suitable mirror planes. Honeycomb is constructed kaleidoscopically: a concretization of Leibniz's monadology is in question. Coxeter group and therefore also the honeycomb is characterized by Coxeter diagram [1] (https://cutt.ly/SjdEZiH) having as its nodes the mirrors and connected by edges labelled by the dihedral angles $\phi = \pi/n$ between the mirror planes. The value of n is written explicitly to the diagram except when it is the minimal value $n = 3$. For instance, the sequence $[(5,3,3,3,3)]$ characterizing tetrahedral-icosahedral honeycomb in H^3 tells that the dihedral angles between the 5 mirror planes are $(\pi/5, \pi/3, \pi/3, \pi/3, \pi/3)$.

Consider now honeycombs in hyperbolic space H^3.

1. The simplest tessellations - regular honeycombs - of H^3 consist of icosahedra and dodecahedra having the same isometry group. That 3 of the 4 most symmetric honeycombs in H^3 have icosahedral symmetries whereas the fourth has cubic symmetries, is a highly encouraging sign. These 4 regular honeycombs are icosahedral honeycomb $\{3,5,3\}$ with 3 icosahedrons meeting along edge; order-5-cubic honeycomb $\{4,3,5\}$ with 5 cubes (rather than 4 as in E^3) meeting along a given edge; and dodecahedral honeycombs of order 4 (5) with 4 (5) dodecahedra meeting along edge. In all these cases the sum of the dihedral angles is larger than 360 ° so that the negative curvature of H^3 is essential for the existence of these honeycombs.

2. What about the combinations of Platonic solids having triangles as faces - tetrahedron, octahedron, and icosahedron? From Wikipedia article [6] (https://cutt.ly/cjaheWC) one learns that there exists honeycombs of H^3 characterized by Schläfli symbol $\{(3,3,5,3)\}$ and Coxeter group with symbol $[(5,3,3,3)]$ consisting of reflections and generating the honeycomb. The regular honeycombs are characterized by 3 integers (say $(3,5,3)$) and the meaning of the code is not quite clear to me but must reflecs the fact that the honeycomb is semiregular.

   Tetrahedron corresponds to (3,3) and icosahedron to (3,5) and octahedron (3,4) as a rectified tetrahedron obtained by contracting edges to their middle points and expanding vertices to faces, has symbol $r(3,3)$. Perhaps (3,3) in (3,3,5,3) refers tCoxetergroupo both tetrahedron and its rectification and (3,5) in (3,3,5,3) to icosahedron. The last "3" tells that 3 identical solid icosahedra, tetrahedra, or octahedra meet at given edge.

   In particular, the tetrahedral-icosahedral honeycomb (TIH) is a compact uniform but not a regular honeycomb, having icosahedra, tetrahedra, and octahedra, all of which have triangular faces, as analogs of unit cells [4, 3, 7] (see https://cutt.ly/xhBwTph, https://cutt.ly/1hBwPRc, and https://cutt.ly/OhBwUOO)). The Wikipedia article [6] contains beautiful illustrations of these honeycombs.

   One can wonder why "tetrahedral-icosahedral honeycomb" does not involve octahedron. This is said to reflect the fact that only tetrahedral and icosahedral cells of the tessellation are regular 3-cells. All these polyhedra are regular as Platonic solids, and it remains unclear to me what the lacking regularity of the octahedron as 3-cell means in the recent context.

For TIH $\{(3,3,5,3\}$ the vertex figure is rhombicosidodecahedron (RID) [9] (https://cutt.ly/yjahitS) discovered already by Kepler. Kepler talked about Harmonices Mundi and I cannot but smile as I recall how I read as a young man a book having fun with Kepler's medieval belief on celestial harmonies and laughed also! Maybe the celestial harmonies are making a glorious comeback!

RID is an Archimedean solid [8] (https://cutt.ly/njahaGN) having 60 vertices corresponding to 12 disjoint pentagons and 20 disjoint triangles with 60 vertices both. RID has as faces 20 triangles assignable to icosahedron, 12 pentagons assignable to dodecahedron plus 30 squares - 62 faces altogether. RID is obtained by radially scaling the distance of icosahedral and dodecahedral faces from origin but keeping the area of the spherical faces the same: this yields squares as additional faces. Triangles and pentagons have only squares as edge neighbors.

Edge figure tells the number of edges meeting at given edge. For TIH it is 3. Regular and single-ringed Coxeter diagram uniform polytopes to which also TIH belongs have a single edge type. Therefore icosahedron, tetrahedron, and octahedron must meet at given edge. That vertex figure contains 3 types of faces (triangles, and squares, and pentagons) presumably reflects this. Recall that the sum of the dihedral angles of T,O, and I is 348.19 °.

One can try to build a more concrete picture about how the Platonic solids are glued together along their triangular faces in the icosahedral-tetrahedral honeycomb.

1. Must to make this concrete, one can regard Platonic solid as a kind of mini Earth with two other Platonic solids glued to its surface like mountains. In all cases one has Platonic analog of a planar lattice of triangles at this mini Earth. To minimize typing call the 3 different Platonic solids T, O, and I.

2. Due to the symmetries one expects that for O and I the triangles correspond to different Platonic solids if they are edge neighbors. For T this is not possible since all faces are edge neighbours. All 6 2+2 configurations of B and C are however related by a rotational symmetry. This already gives a rather satisfactory picture about what the situation looks like at the surface of each mini Earth (I cannot avoid the analogy with inner planets, the living Earth as the largest one would correspond to I!).

3. The radius R of circumscribed inner or or outer sphere gives an idea about the size scales of these Platonic solids when the edge length a is the same for them as it is in the recent case. The following gives the radii of the outer sphere.

$$\begin{array}{ll} \text{tetrahedron} & \frac{R_{T,out}}{a} = \sqrt{\frac{1}{2}} \, , \qquad \frac{R_{T,in}}{a} = \sqrt{\frac{1}{24}} \\ \text{octahedron} & \frac{R_{O,out}}{a} = \sqrt{\frac{3}{4}} \, , \qquad \frac{R_{O,in}}{a} = \sqrt{\frac{1}{6}} \, , \\ \text{icosahedron} & \frac{R_{I,out}}{a} = \frac{1}{2}\sqrt{\phi\sqrt{5}} \, , \phi = \frac{(1+\sqrt{5})}{2} \, , \quad \frac{R_{I,in}}{a} = \frac{\sqrt{3}}{12}\sqrt{3+\sqrt{5}} \, . \end{array} \qquad (2.1)$$

4. The ratios of the outer radii are given by $R_{I,out} : R_{O,out} : R_{T,out} = \sqrt{\phi\sqrt{5}} : \sqrt{\frac{3}{4}} : \sqrt{\frac{1}{2}} \simeq 1.9021 : 0.8660 : 0.7071$. The ratios of the inner radii are given by $R_{I,in} : R_{O,in} : R_{T,in} = \sqrt{\phi\sqrt{5}} : \sqrt{\frac{3}{4}} : \sqrt{\frac{1}{2}} \simeq .756 : 0.408 : 0.2041$. That icosahedron has the largest size, is natural since the total solid angle defined as a sum of the solid angles of the 20 triangles is $4/pi$ and the contribution of an individual triangle is smallest for I and largest for the 4 triangles of T.

2.1.2 Could TIH allow to unify the models of genetic code?

Does this picture help to say anything interesting about the model of bio-harmony and even to unify the models of genetic code?

1. Tesselations define in a natural manner discretizations of MB defining cognitive represenations suggested to relate to the geometric representations for the states of the brain at MB and more generally, for the states of various parts of the biological body at MB. There is evidence for an effective hyperbolic geometry of brain realized in a statistical sense [?]hyperbolic (http://tinyurl.com/ybghux6d) : functionally similar neurons are near to each other in this effective hyperbolic geometry. This evidence is discussed from TGD point of view in [21]: one ends up with a proposal that the MB of the brain provides a geometric representation for the statistical aspects of the brain - kind of abstraction? Information from the brain would be sent by dark Josephson radiation from similar neurons to positions of MB near to each other. This model could generalize to other parts of organism. MBs could form a kind of abstraction hierarchy representing more and more abstract data about the state of organism.

2. Could the icosahedral-tetrahedral tesselation allow a justification for the fusion of 3 icosahedral harmonies with the tetrahedral harmony? Why does the octahedral harmony disappear? Octahedral harmony would mean 6 additional notes assignable to the vertices of octahedron and 8 3-chords and this does not fit with facts.

   Remark: In the Wikipedia article about TIH it is said that octahedrons of TIH are not regular, unfortunately in the sense that I do not understand. Note also that tetrahedral and octahedral harmonies are unique because there is only a single Hamiltonian cycle.

3. Geometrically the tesselation means identification of the neighbouring faces, which gives a justification for the strange looking proposal of gluing tetrahedron to icosahedron in order to fuse 3 icosahedral and one tetrahedral harmony. If also the 3-chords associated with the faces are identified, one can ask whether only icosahedral and tetrahedral harmonies are needed and the chords of the octahedral harmony are determined by them.

   2 3-chords of tetrahedral harmony are the same as those for icosahedral harmony but the 2 3-chords associated with the 2 T-O faces are independent. This would give 62 independent chords (amusingly, 62 happens to be the number of faces of RID).

   One of the tetrahedral chords is necessary since purely icosahedral harmony allows to get only 19 amino-acids identified as the orbits of the chords under the symmetries of a particular icosahedral harmony with 20 chords: one additional chord is needed for the missing amino-acid. Since two icosahedral triangles facing the tetrahedron "eat" 2 further tetrahedral chords, this leaves 1 tetrahedral chord from 4: 3 chords as tetrahedral codons are missing. Could the 3 missing tetrahedral 3-chords correspond to the ordinary DNA codons acting as stop codons? Could the stop codons lack a representation as dark photon triplets or could their frequencies be such that they do not allow 3-resonance with any tRNA?

4. How genes would be realized in the tesselation? Could dark genes correspond to flux tubes forming 1-D sub-tesselations of H^3 induced to the flux tubes? Could gene correspond to a sequence of icosahedron-tetrahedron pairs such that neighboring codons are associated with icosahedron-tetrahedron pairs as cell-neighbors. Two subsequent icosahedrons would have a tetrahedron between them.

   Could the tesselation induced from H^3 to MB be dynamical involving an "activation" of a particular triangle as a codon inside each icosahedron and tetrahedron? Could dark genes at the flux tubes have these codons as induced dark codon sequences? Could "activation" mean that the triangle representing particular codon is accompanied by 3 dark protons at its vertices and representing the same genetic codon? The representations in terms of dark protons triplets, as triangles of icosahedron and tetrahedron, and as dark photon triplets would fuse to single representation. There could be a representation also for stop codons in terms of 3 dark protons but there would not be no triangle where to locate them so that coding would stop! The missing dark codon would signify the end of the gene.

This would give the long-sought connection between dark codons realized as dark triplets and dark codons realizing bio-harmony and dark codons realized as dark photon triplets generated in the cyclotron transitions of dark codons. An essential role would be played by Galois confinement [22] stating that these triplets behave like dynamical units - just like 3 confined quarks forming a baryon. Galois confinement generalizes to the level of genes.

5. This proposal is of course one of the many variations of single theme developed during years. What is new that the proposal would make the roles of the icosahedral and tetrahedral geometries concrete, not at the level of bio-molecules but at the level of their MBs. A profound dramatic generalization of the notion of genetic code from biology to the level of fundamental physics is also suggestive. Even a hierarchy of genetic codes in various scales can be considered.

The interpretation of various harmonies as correlates of emotions implies that each icosahedral-tetrahedral unit of the tesselation would have its own varying emotional state expressed and affected by biochemical level via different interaction actions with ordinary biomatter realized in terms of dark photon N-resonance with targets depending on the emotional state [24, 25, 26]. This could serve as a universal mechanism of bio-control by MB applying also to epigenesis.

There are still several open questions: in particular, what is the deeper reason for the fusion of just 3 icosahedral bio-harmonies. That the number of the dark codons is 64 is a partial reason but is this enough.

6. There are reasons to ask whether the cell membrane and microtubuli could provide a 2-D realizations of the genetic code [24]. If genes are induced as 1-D sub-tesselations from that of MB, there is no reason to exclude 2-D or even 3-D induced tesselations.

7. I cannot avoid the temptation of mentioning the notion of memetic code [12], which was my first idea about genetic code and proposed as a generalization of genetic code by starting from a speculated hierarchy of Mersenne primes, whose members would come as $M(n + 1) = M_{M(n)}$, $M_n = 2^n - 1$, ($M(2) = 2$. This gives the Mersenne primes $M(2) = M_2 = 3$, $M(3) = 2^3 - 1 = 7$, $M(4) = M_7 = 2^7 - 1$, $M(5) = M_{127} = 2^{127-1}$. It is not known whether the hierarchy continues. M_7 would correspond to the ordinary genetic code and M_{127} to memetic code with codons realizable as sequences of 20 codons.

Could memetic code be realized by TIH? Could one consider a planar or cylindrical sub-tesselation with a width of 20 tetrahedral-icosahedral pairs? If the size assignable to single pair is that of DNA codon - 1 nm roughly - the width would be about 20 nm which might relate to the radial scale of the microbutubuli.

Acknowledgements: I am grateful for Reza Rastmanesh for generous help in the preparation of the manuscript.

Received January 19, 2021; Accepted October 1, 2021

References

[1] Coxeter-Dynkin diagram. Available at: https://en.wikipedia.org/wiki/Coxeter-_Dynkin_diagram.

[2] Coxeter group. Available at: https://en.wikipedia.org/wiki/Coxeter_group.

[3] Honeycomb geometry. Available at: https://en.wikipedia.org/wiki/Honeycomb_(geometry).

[4] List of regular polytopes and compounds. Available at: https://en.wikipedia.org/wiki/List_of_regular_polytopes_and_compounds.

[5] Schläfli symbol. Available at: http://en.wikipedia.org/wiki/Schläfli_symbol.

[6] Tetrahedral-icosahedral honeycomb. Available at: https://en.wikipedia.org/wiki/Tetrahedral-icosahedral_honeycomb.

[7] Uniform honeycomes in hyperbolic space. Available at: https://en.wikipedia.org/wiki/Uniform_honeycombs_in_hyperbolic_space.

[8] Archimedean solids. Available at: https://en.wikipedia.org/wiki/Archimedean_solids.

[9] Rhombicosidodecahedron. Available at: https://en.wikipedia.org/wiki/Rhombicosidodecahedron.

[10] Vertex figure. Available at: https://en.wikipedia.org/wiki/Vertex_figure.

[11] Cacciola A et al. Coalescent embedding in the hyperbolic space unsupervisedly discloses the hidden geometry of the brain, 2017. Available at:https://arxiv.org/pdf/1705.04192.pdf.

[12] Pitkänen M. Genes and Memes. In *Genes and Memes: Part I*. Available at: http://tgdtheory.fi/pdfpool/genememec.pdf, 2006.

[13] Pitkänen M. Geometric theory of harmony. Available at: http://tgdtheory.fi/public_html/articles/harmonytheory.pdf., 2014.

[14] Pitkänen M. Philosophy of Adelic Physics. In *Trends and Mathematical Methods in Interdisciplinary Mathematical Sciences*, pages 241–319. Springer.Available at: https://link.springer.com/chapter/10.1007/978-3-319-55612-3_11, 2017.

[15] Pitkänen M. What does cognitive representability really mean? Available at: http://tgdtheory.fi/public_html/articles/cognrepres.pdf., 2017.

[16] Pitkänen M. An overall view about models of genetic code and bio-harmony. Available at: http://tgdtheory.fi/public_html/articles/gcharm.pdf., 2019.

[17] Pitkänen M. Scattering amplitudes and orbits of cognitive representations under subgroup of symplectic group respecting the extension of rationals . Available at: http://tgdtheory.fi/public_html/articles/symplorbsm.pdf., 2019.

[18] Pitkänen M. When do partonic 2-surfaces and string world sheets define large cognitive representations? Available at: http://tgdtheory.fi/public_html/articles/elliptic.pdf., 2019.

[19] Pitkänen M. A critical re-examination of $M^8 - H$ duality hypothesis: part I. Available at: http://tgdtheory.fi/public_html/articles/M8H1.pdf., 2020.

[20] Pitkänen M. A critical re-examination of $M^8 - H$ duality hypothesis: part II. Available at: http://tgdtheory.fi/public_html/articles/M82.pdf., 2020.

[21] Pitkänen M. Could brain be represented as a hyperbolic geometry? Available at: `http://tgdtheory.fi/public_html/articles/hyperbolicbrain.pdf.`, 2020.

[22] Pitkänen M. How to compose beautiful music of light in bio-harmony? Research Gate: `https://www.researchgate.net/publication/344623253_How_to_compose_beautiful_music_of_light_in_bio-harmony.`, 2020.

[23] Pitkänen M. The dynamics of SSFRs as quantum measurement cascades in the group algebra of Galois group. Available at: `http://tgdtheory.fi/public_html/articles/SSFRGalois.pdf.`, 2020.

[24] Pitkänen M and Rastmanesh R. The based view about dark matter at the level of molecular biology. Available at: `http://tgdtheory.fi/public_html/articles/darkchemi.pdf.`, 2020.

[25] Pitkänen M and Rastmanesh R. New Physics View about Language: part I. Available at: `http://tgdtheory.fi/public_html/articles/languageTGD1.pdf.`, 2020.

[26] Pitkänen M and Rastmanesh R. New Physics View about Language: part II. Available at: `http://tgdtheory.fi/public_html/articles/languageTGD2.pdf.`, 2020.

Homeostasis as Self-organized Quantum Criticality

M. Pitkänen[1] and R. Rastmanesh[2,3]

[1]Independent researcher. [1]
[2]Member of The Nutrition Society, London, UK.
[3]Member of The American Physical Society, USA.

Abstract

Cold shock proteins (CSPs) and heat shock proteins (HSPs) have a great deal of similarity and have much more general functions, so it is easier to talk about stress proteins (SPs) having two different modes of operation. The attempt to understand various functions of SPs led to much more general problem: how self-organized quantum criticality (SOQC) is possible? Criticality means by definition instability but SOQC is stable, which seems to be in conflict with the standard thermodynamics. In fact, living systems as a whole are quantum critical and manage to stay near quantum criticality, which means SOQC. This is nothing but homeostasis usually understood as a complex control system needed to keep living systems in flow equilibrium. Zero energy ontology (ZEO) forming the basics of TGD (Topological Geometrodynamics) inspired quantum measurement theory extends to a quantum theory of consciousness and living systems and predicts that the arrow of time changes in ordinary ("big") state function reductions. ZEO leads to a theory of quantum self-organization and time reversal means that dissipation in reversed direction looks like extraction of energy from the environment for the observer with standard time direction. The change of the arrow of time transforms critical states from repellers to attractors and makes SOQC possible. SOQC and homeostasis would result automatically rather than being forced. Magnetic body (MB) is another key notion. MB has a maximal temperature known as Hagedorn temperature T_H crucial for understanding SOQC and functioning of SPs. T_H would relate closely to physiological temperature in biomatter.

1 Introduction

This article started as an attempt to understand the properties of cold shock proteins (CSPs) and heat shock proteins (HSPs) in TGD framework. As a matter of fact , these proteins have great deal of similarity and have much more general functions, so it is easier to talk about stress proteins (SPs) having two different modes of operation. time As we proceed, it will be revealed that this issue is only one particular facet of a much bigger problem: how self-organized quantum criticality (SOQC) is possible? Criticality means by definition instability but SOQC is stable, which seems to be in conflict with the standard thermodynamics. In fact, living systems as a whole seem to be quantum critical [?] and manage to stay near criticality, which means SOQC. Note that the self-organized criticality (SOC) is generalized to SOQC.

Topological Geometrodynamics (TGD) [?] [?, ?] is a 43 year old proposal for a unification of fundamental interactions. Zero energy ontology (ZEO) [?] is basic aspect of quantum TGD and allows to extend quantum measurement theory to a theory of consciousness and of living systems. ZEO also leads to a quantum theory of self-organization [?] predicting both arrows of time. Could ZEO make SOQC possible as well?

---

[1]Correspondence: Matti Pitkänen http://tgdtheory.com/. Address: Rinnekatu 2-4 A8, 03620, Karkkila, Finland. Email: matpitka6@gmail.com. Email: matpitka6@gmail.com.

1.1 Summary of the basic properties of CSPs and HSPs

Let's consider a summary of CSPs and HSPs or briefly SPs.

1. There is a large variety of cold shock proteins (CSP) and heat shock proteins (HSPs). CSPs and HSPs are essentially the same proteins and labelled by HSPX, where X denotes the molecular weight of the protein in kDaltons. The value range of X includes the values $\{22, 60, 70, 90, 104, 110\}$ and HSPs are classified into 6 families: small HSPs, HSPX, $X \in \{40, 60, 70, 90, 110\}$. At least HSP70 [?] and HSP90 [?] have ATPase at their end whereas HSP60 has ATP binding site [?]. CSPs and HSPs consist of about $10^3 - 10^4$ amino acids so that X varies by one order of magnitude.

    Their lengths in the un-folded active configuration are below 1 micrometer. CSPs/HSPs [?, ?, ?, ?] are expressed when the temperature of the organism is reduced /increased from the physiological temperature. CSPs possess cold-shock domains [?] consisting of about 70-80 amino-acids thought to be crucial for their function. Part of the domain is similar to the so called RNP-1 RNA-binding motif. In fact, it has turned that CSP and HSP are essentially the same object and stress protein (SP) is a more appropriate term.

    Wikipedia article about cold shock domain [?] mentions Escherichia Coli as an example. When the temperature is reduced from 37 °C to 10 °C, there is 4-5 hours lag phase after which growth is resumed at a reduced rate. During lag phase expression of around 13 proteins containing cold shock domains is increased 2-10 fold. CSPs are thought to help the cell to survive in temperatures lower than optimum growth temperature, by contrast with HSPs, which help the cell to survive in temperatures greater than the optimum, possibly by condensation of the chromosome and organization of the prokaryotic nucleoid. What is the mechanism behinds SP property is the main question.

2. SPs have a multitude of functions involved with the regulation, maintenance and healing of the system [?, ?, ?, ?, ?]. They appear in stress situations like starvation, exposure to cold or heat or to UV light, during wound healing or tissue remodeling, and during the development of the embryo. SPs can act as chaperones [?] and as ATPAses [?, ?].

    SPs facilitate translation, and protein folding in these situations, which suggests that they are able to induce local heating/cooling of the molecules involved in these processes. CSPs could be considered like ovens and HSPs like coolants; systems with very large heat capacity acting as a heat bath and therefore able to perform temperature control. SPs serve as kind of molecular blacksmiths - or technical staff - stabilizing new proteins to facilitate correct folding and helping to refold damaged proteins. The blacksmith analogy suggests that this involves a local "melting" of proteins making it possible to modify them.

    What "melting" could mean in this context? One can distinguish between denaturation in which the folding ability is not lost and melting in which it is lost. Either local denaturation or even melting would be involved depending on how large the temperature increase is. In a aqueous environment the melting of water surrounding the protein as splitting of hydrogen bonds is also involved. One could also speak also about local unfolding of protein.

3. There is evidence for large change ΔC_p of heat capacity C_p ($C_p = dE/dT$ for pressure changing feed of heat energy) for formation ion nucleotide-CSP fusion [?]. This could be due to the high C_p of CSP. The value of heat capacity of SPs could be large only *in vivo*, not *in vitro*.

4. HSPs can appear even in hyper-thermophiles living in very hot places. This suggests that CSPs and HSPs are basically identical - more or less - but operate in different modes. CSPs must be able to extract metabolic energy and they indeed act as ATPases. HSPs must be able to extract thermal energy. If they are able to change their arrow of time as ZEO suggests, they can do this by dissipating with a reversed arrow of time.

    To elucidate the topic from other angles, the following key questions should be answered:

1. Are CSPs and HSPs essentially identical?

2. Can one assign to SPs a high heat capacity (HHC) possibly explaining their ability to regulate temperature by acting as a heat bath? One can also ask whether HHC is present only *in vivo* that is in a aqueous environment and whether it is present only in the unfolded configuration of HP?

1.2 The notion of quantum criticality

The basic postulate of quantum TGD is that the TGD Universe is quantum critical [?, ?] [?, ?]. There is only a single parameter, Kähler coupling strength α_K mathematically analogous to a temperature and theory is unique by requiring that it is analogous to critical temperature. Kähler coupling strength has discrete spectrum labelled by the parameters of the extensions of rationals. Discrete p-adic coupling constant evolution replacing continuous coupling constant evolution is one aspect of quantum criticality.

What does quantum criticality mean?

1. Quite generally, critical states define higher-dimensional surfaces in the space of states labelled for instance by thermo-dynamical parameters like temperature, pressure, volume, and chemical potentials. Critical lines in the (P,T) plane is one example. Bringing in more variables one gets critical 2-surfaces, 3-surfaces, etc. For instance, in Thom's catastrophe theory [?] cusp catastrophe corresponds to a V-shaped line, whose vertex is a critical point whereas butterflly catasrophe to 2-D critical surface. In thermodynamics the presence of additional thermodynamical variables like magnetization besides P and T leads to higher-dimensional critical surfaces.

2. There is a hierarchy of criticalities: there are criticalities inside criticalities. Critical point is the highest form of criticality for finite-D systems. Triple point, for instance, for water in which one cannot tell whether the phase is solid, liquid or gas. This applies completely generally irrespective of whether the system is a thermo-dynamical or quantal system. Also the catastrophe theory of Thom gives the same picture [?]. The catastrophe graphs available in the Wikipedia article illustrate the situation for lower-dimensional catastrophes.

3. In TGD framework finite measurement resolution implies that the number of degrees of freedom (DFs) is effectively finite. Quantum criticality with finite measurement resolution is realized as an infinite number of hierarchies of inclusions of extensions of rationals. They correspond to inclusion hierarchies of hyperfinite factors of type II_1 (HFFs). The included HFF defines the DFs remaining below measurement resolution and it is possible to assign to the detected DFs dynamical symmetry groups, which are finite-dimensional. The symmetry group in never reachable ideal measurement resolution is infinite-D super-symplectic group of isometries of "world of classical worlds" (WCW) consisting of preferred extremals of Kähler action as analogs of Bohr orbits. Super-symplectic group extends the symmetries of superstring models [?] [?, ?, ?, ?].

4. Criticality in living systems is a special case of criticality - and as the work of Kauffman [?] suggests - of quantum crticality as well. Living matter as we know, it most probably corresponds to extremely high level of criticality so that very many variables are nearly critical, not only temperature but also pressure. This relates directly to the high value of h_{eff} serving as IQ. The higher the value of h_{eff}, the higher the complexity of the system, and the larger the fluctuations and the scale of quantum coherence. There is a fractal hierarchy of increasingly quantum critical systems labelled by a hierarchy of increasing scales (also time scales).

   In ZEO classical physics is an exact part of quantum physics and quantum physics prevails in all scales. ZEO makes discontinuous macroscopic BSFRs to look like smooth deterministic time evolutions for the external observer with opposite arrow of time so that the illusion that physics is classical in long length scales is created.

Number theoretical physics or adelic physics [?] is the cornerstone of TGD inspired theory of cognition and living matter and makes powerful predictions.

p-Adic length scale hypothesis deserves to be mentioned as an example of prediction since it has direct relevance for SPs.

1. p-Adic length scale hypothesis predicts that preferred p-adic length scales correspond to primes $p \simeq 2^k$: $L(k) = 2^{(k-151)/2}L(151)$, $L(151) \simeq 10$ nm, thickness of neuronal membrane and a scale often appearing molecular biology.

2. TGD predicts 4 especially interesting p-adic length scales in the range 10 nm- 25 μ. One could speak of a number theoretical miracle. They correspond to Gaussian Mersenne primes $M_{G,k} = (1+i)^{k-1}$ with prime $k \in \{151, 157, 163, 167\}$ and could define fundamental scales related with DNA coiling for instance.

3. The p-adic length scale $L(k=167) = 2^{(167-151)/2}L(151) = 2.5$ μ m so that SPs could correspond to $k \in \{165, 167, 169\}$. $L(167)$ corresponds to the largest Gaussian Mersenne in the above series of 4 Gaussian Mersennes and to the size of cell nucleus. The size scale of a cold shock domain in turn corresponds to $L(157)$, also associated with Gaussian Mersenne. Note that the wavelength defined by $L(167)$ corresponds rather precisely to the metabolic currency .5 eV.

4. HSPX, $X \in \{60, 70, 90\}$ corresponds to a mass of X kDaltons (Dalton corresponds to proton mass). From the average mass 110 Dalton of amino acid and length of 1 nm one deduces that the straight HSP60, HSP70, and HSP90 have lengths about .55 μm, .64 μ, and .8 μm. The proportionality of the protein mass to length suggests that the energy scale assignable to HSPX is proportional to X. (HSP60, HSP70, HSP90) would have energy scales (2.27, 1.95,1.5 eV) for $h_{eff} = h$ naturally assignable to biomolecules. The lower boundary of visible photon energies is a 1.7 eV.

   Remark: One has $h = h_{eff} = nh_0$ for $n = 6$. What if one assumes $n = 2$ giving $h_{eff} = h/3$ for which the observations of Randel Mills [?] give support [?]? This scales down the energy scales by factor 1/3 to (.77,.65,0.5) eV not far from the nominal value of metabolic energy currency of about .5 eV.

   There are strong motivations to assign to HSPs the thermal energy $E = T = .031$ eV at physiological temperature: this is not the energy $E_{max} = .084$ eV at the maximum of the energy distribution, which is by a factor 2.82 higher than E. The energies above are however larger by more than one order of magnitude. This scale should be assigned with the MBs of SPs.

5. The wavelengths assignable to HSPs correspond to the "notes" represented by dark photon frequencies. There is an amusing co-incidence suggesting a connection with the model of bio-harmony [?, ?]: the ratios of energy scales of HSP60 and HSP70 to the HSP90 energy are 3/2 and 1.3, respectively. If HSP90 corresponds to note C, HSP60 corresponds to G and HSP70 to note E with ratio 1.33. This gives C major chord in a reasonable approximation! Probably this is an accident. Note also that the weights X of HSPXs are only nominal values.

1.3 Hagedorn temperature, HHC, and self-organized quantum criticality (SOC)

Self-organized criticality (SOC) is an empirically verified notion. For instance, sand piles are SOQC systems. The paradoxical property of SOQC is that although criticality suggests instability, these systems stay around criticality. In standard physics SOQC is not well-understood. TGD based model for SOQC involves two basic elements: ZEO and Hagedorn temperature.

1. ZEO predicts that quantum coherence is possible in all scales due to the hierarchy of effective Planck constants predicted by adelic physics. "Big" (ordinary) state function reductions (BSFRs) change the arrow of time [?]. Dissipation in reversed arrow of time looks like generation of order and

structures instead of their decay - that is self-organization. Hence SOQC could be made possible by the instability of quantum critical systems in non-standard time direction. The system paradoxically attracted by the critical manifold in standard time direction would be repelled from it in an opposite time direction as criticality indeed requires.

2. Surfaces are systems with infinite number of DFs. Strings satisfy this condition as also magnetic flux tubes idealizable as strings in reasonable approximation. The number of DFs is infinite and this implies that when one heats this kind of system, the temperature grows slowly since heat energy excites new DFs. The system's maximum temperature is known as Hagedorn temperature and it depends on string tension for strings.

In the TGD framework, magnetic flux tubes can be approximated as strings characterized by a string tension decreasing in long p-adic length scales. This implies a very high value of heat capacity since very small change of temperature implies very large flow of energy between the system and environment.

T_H could be a general property of MB in all scales (this does not yet imply SOQC property). An entire hierarchy of Hagedorn temperatures determined by the string tension of the flux tube, and naturally identifiable as critical temperatures is predicted. The temperature is equal to the thermal energy of massless excitations such as photons emitted by the flux tube modellable as a black body.

Remark: If the condition $h_{eff} = h_{gr}$ [?], where h_{gr} is gravitational Planck constant introduced originally by Nottale [?], holds true, the cyclotron energies of the dark photons do not depend on h_{eff}, which makes them an ideal tool of quantum control.

Hagedorn temperature would make them SOQC systems by temperature regulation if CSP type systems are present they can serve as ovens by liberating heat energy and force the local temperature of environment to their own temperature near T_H. Their own temperature is reduced very little in the process. These systems can also act as HSP/CSP type systems by extracting heat energy from/providing it to the environment and in this manner reduce/increase the local temperature. System would be able to regulate its temperature.

A natural hypothesis is that T_H corresponds to quantum critical temperature and in living matter to the physiological temperature. The ability to regulate the local temperature so that it stays near T_H has interpretation as self-organized (quantum) criticality (SOC). In the TGD framework these notions are more or less equivalent since classical physics is an exact part of quantum physics and BSFRs create the illusion that the Universe is classical in long (actually all!) scales.

Homeostasis is a basic aspect of living systems. System tends to preserve its flow equilibrium and opposes the attempts to modify it. Homeostasis involves complex many-levels field back circuits involving excitatory and inhibitory elements. If living systems are indeed quantum critical systems, homeostasis could more or less reduce to SOQC as a basic property of the TGD Universe.

2 The basic ideas about SPs

The TGD based model for SPs relies on the notion of MB carrying dark matter as $h_{eff} > h$ phases and the notions of heat transfer and heat capacity. The basic idea is that at least in aqueous environment the MBs of biomolecules in general have a large number of DFs and act as heat reservoirs with a stable temperature near a Hagedorn temperature. MBs of SPs have also high heat transfer rates between the thermal environment of the ordinary matter. ZEO - in particular time reversal - makes it possible to realize thermal regulation in terms of SOQC. On the other hand, information carrying biomolecules cannot have high heat transfer rate with environment.

2.1 Conditions on the heat transfer rates between the systems involved

To avoid lengthy explanations, it is appropriate to introduce some shorthand notations. Denote by $j_H(X-Y)$ heat transfer rate between systems X and Y. Denote by E. Denote BB(X) the biological body of system X. X can denote the ordinary biomolecule (DNA,RNA,protein) denoted by BM or stress protein SP.

There are several conditions on the model explaining the HHC

1. $j_H(MB(SP)-E)$ should be high so that the MB of SP can rapidly adapt to temperature changes and extract thermal energy from the environment and act as an oven or a coolant. $j_H(MB(SP)-BM$ should be high so that CSPs could rapidly warm up BMs for processes like translation, transcription and folding.

    $j_H(MB(SP)-BM$ can be also high if heat transfer occurs indirectly via $MB(BM)$. This requires that both $j_H(MB(BM)-BM$ and $j_H(MB(SP)-MB(BM)$ are high. However, the large value of $j_H(MB(BM)-BM$ implies that BMs can take care of temperature regulation without the help of SPs. Hence this option does not seem to be consistent with empirical facts. Hence $j_H(MB(BM)-BM$ must be low.

    There is also a deeper rational for this. The MBs of ordinary bio-molecules must carry information and cannot be thermalized so that the energy transfer rate between them and their BB and between them and the environment must be low.

2. In CSP mode the MBs of SPs should actively extract energy from fats. The BMs should extract thermal energy from MBs of SPs. In HSP mode MBs of SPs at temperature than that of the local thermal environment (including BMs) should cool it by absorbing thermal energy from it.

The following table summarizes the constraints on the symmetric matrix of heat transfer rates $j_H(A, B)$ for various combinations of subsystems X and Y. The shorthand notations are (SP, BM, E) for (stress protein, basic biomolecule, environment) and $MB(X)$ for the MB of molecule X. Environment E is taken as the thermal environment at the level of ordinary matter. The diagonal heat transfer rates are not considered. H/L for the matrix element $j_H(X,Y)$ of the table means that its value can be large/small. The symbol "*" means that this particular transfer is not relevant.

$$
\begin{array}{cccccc}
X/Y & SP & MB(SP) & BM & MB(BM) & E \\
SP & * & H & * & * & * \\
MB(SP) & H & * & H & * & H \\
BM & * & H & * & L & * \\
MB(BM) & * & * & L & * & * \\
E & * & H & * & * & *
\end{array}
\quad (2.1)
$$

In the minimal scenario the only constraints are on $j_J(SP, MB(SP)$ (H), $j_J(BM, MB(SP)$ (H), and $j_J(BM, MB(BM)$ (L).

The natural question is what makes it possible for the MBs of SPs to gain energy.

1. The first manner to get energy is heat transfer from the environment. Passive heat transfer would involve either ordinary photons transformed to dark photons and absorbed by MB(SP) or active heat extraction in time reversed mode involving emission of dark photons transformed to ordinary photons and absorbed by ordinary matter. The energies should be in the range of thermal energies at physiological temperatures.

2. The negative energy photons from the MB of biomolecule can be also received by other MBs acting as analogs of population reversed laser. Thermalisation is expected to occur if there is large number of this kind of states. MB should allow almost continuum of cyclotron energy state in the energy resolution defined by the size scale of the molecules.

3. At least some SPs such as HSP70 and HSP90 could act as ATPases providing the heat energy at their MBs to drive ADP → ATP process. They would act as general purpose quantum heat engines with MB acting as a heat bath running the ATPase machinery. Heat engine function requires a heating of the MB SP to a temperature above the local physiological temperature but below the Hagedorn temperature: in ZEO time reversal for the MB of SP allows this: it would look like extraction of thermal energy from the environment. Part of the energy heating MB of SP could come from the binding of ATP to ATPAse part of PS. This energy is in the range of 3-7 eV for nucleotides and could heat the MB of SP.

One could also consider remote metabolism for the molecules receiving the metabolic energy quantum with a negative energy photon inducing $ATP \to ADP$. Note that the metabolic energy quantum .5 eV is in infra-rede (IR) range and corresponds to 2.4 μm wavelength very near to the largest p-adic length scale $L(167)$ in the quadruplet of primes $k \in \{151, 157, 163, 167\}$ defining four Gaussian Mersennes and defining the size scale of nucleus.

Now, consider the extraction of heat energy from the environment:

1. The energies assignable to the photon wavelengths defined by the lengths of HSPX proteins are proportional to $1/X$ and above 1.5 eV, which is considerably above the energy of thermal photon at the maximum of Planck distribution for energy is $E_{max} = .084$ eV).

2. The energy transfer would be based on energy resonance and is possible only if the cyclotron frequency spectrum of dark particles contains energies possessed by molecules in their spectrum in infrared range. This poses a condition on the cyclotron energies $E = \hbar_{eff}eB/m$ assumed to be in bio-photon energy range: this requires that $h_{eff} = nh_0 = \hbar_{gr} = BMm/v_0$ is large: one has $E = GMB/v_0$ does not depend on the mass of charged particle. Cyclotron energies involve also the contribution from a longitudinal motion along the flux tube. The energy scale for dark photon is now $\hbar_{eff}/L$ and also universal since L scales as $\hbar_{gr}$. If L is small the energy scale is so large that longitudinal DFs are not excited and thermalization does not occur. Same is true if B is large enough.

Magnetic field strength is expected to scale like $1/L^2(k)$, where $L(k)$ is the p-adic length scale characterizing the molecule. The endogenous magnetic field $B_{end} = 2B_E/5$ identified as the monopole flux part of the Earth's magnetic field is expected to define an important value in the spectrum of magnetic fields. The corresponding p-adic length scale corresponds to the length scales assignable to SPs. Also octaves of this value are expected and the model of bio-harmony [?, ?] suggests that the preferred values are given by 12-note scale.

For short linear molecules the energy scales would be too high to allow thermalization so that these molecules can serve as information molecules. For long DNA one has length scale hierarchy and thermalization can occur only in long enough length scales. Human DNA has total length of order 1 meter but if the size of DNA defines the p-adic length scale, then DNA does not thermalize since the size of nucleus is not larger than $L(167) = 2.5$ μm. Note that DNA defines a length scale hierarchy in codons, genes, and also coiling scales define hierarchy levels. When the length of the molecules is longer than the wavelength of thermal photon at room temperature, one expects thermalisation to occur. SPs have lengths below 1 μm.

3. The thermalization should take place for the MBs of SPs. There are two energy scales associated with the cyclotron energies and the free motion along the flux tube respectively. Thermal energy scale could correspond to either of these length scales.

   (a) Cyclotron energy scale is given by $E_c = GMB/v_0$ for $h_{eff} = h_{gr}$ and the scales are proportional to B. Longitudinal energy scale dose not depend on h_{eff} since the flux tube length scales like h_{eff}. Since B scales like $1/L^2(k)$, cyclotron length scale increases for small protein sizes. This suggests that thermalization is associated with the cyclotron DF and appears for large enough p-adic length scales characterizig protein size.

(b) Longitudinal energy scale naturally corresponds to the lengh of protein for $h_{eff} = n$. The energy scale of longitudinal excitations is consirably above the thermal energy scale so that thermalization would not be possible. It might be however possible to transfer energy from these DFs to the MB of SP where it is transformed to thermal energy.

2.2 A new physics model for HHC

Now, consider a more concrete new physics model for HHC:

1. HHC suggests the existence of new DFs to which energy is stored so that temperature is not raised as new DFs become available.

2. In the theory of extended objects like strings, the very large number (infinite) of degress of freedom (DFs) implies a maximal temperature T_H known as Hagedorn temperature. Flux tubes are extended objects. This suggests that the MBs of SPs are near to the Hagedorn temperature defining the maximal temperature for their MBs. Also the assumption that the physiological temperature is near but usually below T_H: this condition allows SP to act as heat engine. This cannot be true for the information carrying biomolecules such as DNA, RNA and proteins since thermalization destroys information. Therefore they must have a temperature much below T_H.

3. In a hot environment the existence of Hagedorn temperature T_H for the MB of HSP means that the thermal energy is transferred from the environment to the MB of HSP. This tends to reduce the local temperature of the environment towards T_H. HSP would act as an ideal coolant. Their presence would faciliate the basic functions of cells.

4. CSP and its MB would be at temperature near T_H and could act as an oven. Their presence around DNA, RNA, and proteins would raise their temperature locally and facilitate transcription, translation and protein folding and unfolding otherwise prevented by a low temperature.

5. SPs could act as heat engines providing heat energy to molecular motors [?]. This entails SP to have a temperature higher than the temperature of environment. In ZEO this is possoble by using a time reversed mode for SP to extract energy from the environment. Many SPs have ATPase at their end and this would make them universal heat engines providing the work as metabolic energy currency for any molecular user.

6. Quite generally, by their ATPase property, many SPs could act as metabolic energy sources in stressful situations - this comprises many other situations in addition to low and high temperatures. Metabolic energy feed increases h_{eff} and would increase the scale quantum coherence reduced in the damage of DNA, proteins and tissue, for instance. After this, the system could self-organize to the healed state. For instance, CSPs could induce local melting of misfolded proteins leading to a repair. CSPs act as chaperones and their basic tool would be local "melting" (remind our operational definition of "melting") by feeding heat energy - allowing to establish a correct conformation.

7. The MBs of SPs could extract their thermal energy from the thermal energy of the environment in time reversed mode allowed by ZEP allowing the temperature of SP to even exceed that of environment in the final state of BSFR.

Consider a quantitative estimate.

1. For a typical flux tube length is larger than the radius of the flux tube. The critical temperature identified as Hagedorn temperature corresponds to a typical thermal energy of the flux tube and is determined by flux tube length and its string tension. The critical temperature is inversely proportional to the length of the flux tube.

2. Critical temperature T_H roughly corresponds to the energy of a photon with wavelength equal to the flux tube length L: $E = T_H \sim h_{eff}/L$. For $h_{eff} = h$ the flux tube length corresponds to the length scale of CSP but for large values of $h_{eff} = h_{gr}$ it corresponds to a scale of even Earth. The energies and temperature T_H are however the same irrespective of the value of h_{eff} and thus length of flux tube.

3. The rough estimate is that for physiological temperatures T_{ph} around T_H, the length for $h_{eff} = h$ the wave length for a thermal photon at temperate 310 K the maximum of energy distribution is around 14.7 μm: note that the sizes of most animal and plant cells are oin the rage of 10-100 μm. For the wavelength distribution the wavelength for the maximum is roughly 7 μm. CSPs and HSPs consist of about 100-1000 amino acids or so. Length would be in the range .1-1 μm. The energies of photons with a wave length of straight SP are definitely above thermal energy range.

Some questions are in order.

1. If the new DFs are associated with MB, what can one say about the value of h_{eff} serving as IQ could be? SPs are possessed already by bacteria which suggests that the value of h_{eff} cannot be very large. Acting as a chaperon is a control function, which suggests a higher than normal value of h_{eff}. Higher than normal value ignites intriguing question whether they have higher IQ (as a value of h_{eff} characterizing number theoretic complexity) than other proteins helping to survive in difficult situations. On the other hand, the thermalization means that SP flux tubes cannot carry information unlike the flux tubes of basic bio-molecules with their MBs at very low temperature.

2. Cell membrane must stay flexible as temperature is lowered. This is known to be achieved by a generation of unsaturated bonds to lipids. This involves desaturase enzyme creating C-C double bond. Desaturase enzymes are not SPs. SPs can however faciliate the transcription and translation of desaturase enzymes.

2.3 Physiological temperature as Hagedorn temperature, local temperature regulation, and self organized quantum criticality

The notions of quantum criticality, self-organized quantum criticality (SOC) and Hagedorn temperature leads to a new physics based model for the explanation of SP functions.

1. Hagedorn temperature T_H as a maximal temperature of MB of stress protein would be crucial for its functioning. Why the physiological temperature is around 310 K is one of the puzzles of biology. The work of Kauffman [?] suggests that the interpretation as a quantum critical temperature is appropriate. TGD predicts a hierarchy of quantum critical temperatures. The natural guess would be that this quantum critical temperature is Hagedorn temperature realized at the level of MB asymptotically: in practice, the temperature of MB would be somewhat below T_H.

    This would facilitate temperature regulation or perhaps even make it possible. At quantum criticality also long length scale quantum fluctuations are possible and this makes modifications of the system possible - say damaged proteins. If the temperature T of the environment at BB is above T_H, the thermal energy flows to MB of SP and its temperature T is reduced. MB can also make BSFR reversing the arrow of time and extract thermal energy from the environment.

2. Self-organized criticality (SOC) generalizes to self-organized quantum criticality (SOQC) in the TGD framework. SOC is well-known but it is not understood. For instance, sand piles are SOC systems. They tend to approach a critical state, which looks paradoxical since just the opposite should hold for critical systems by their defining property which makes them unstable! Critical system is optimal for measuring and representing since it has a large number of different states with roughly the same energy. Therefore biosystems should be critical systems.

The basic objection against SOC and SOQC is that SCs are unstable by definition. In ZEO this objection can be circumvented. Quantum coherence is possible in all scales and in BSFRs the arrow of time is changed. This transforms the critical manifold from a repeller to an attractor and time reversals make SOQC possible. The occurrence of SOQC would be direct empirical proof for the ZEO and its most dramatic predictions.

What is the distinction between CSP and HSP modes of SPs? SOQC according to ZEO suggests that time reversal could explain this difference. How do the time reversals for CSP and HSP modes differ? The following picture is suggestive.

1. The time reversal occurs for the MB of SP in HSP mode so that they extract thermal energy from environment.

2. The time reversal occurs for the MBs molecules interacting with SPs in CSP mode so that they can extract heat energy from the MB of CSP.

It has been already told that homeostasis in presence of quantum criticality is essentially quantum critical SOC.

2.4 $\Delta C_p > 0$ for HSP90-nucleotide binding as support for the model

Chistopher et al have studied enthalpy driven reactions involving nucleotide or ansamycin bimnding to HSP90: the title of the article [?] is "Structural–Thermodynamic Relationships of Interactions in the N-Terminal ATP-Binding Domain". These reactions occurring in constant pressure are enthalpy driven meaning that heat is liberated in these reactions - the second option would be entropy driven reaction in which the large entropy gain makes reaction possible. The formation of a bound state means a reduction of DFs suggesting a decrease of the heat capacity C_p of the combined system.

Researchers however find $\Delta C_p > 0$ when another reactant is nucleotide but not for the ansamycin case. Intuitively, the number of DFs should increase to explain this. The authors of the article discussed a number of explanations for their unexpected finding.

The presence of MB means new hidden DFs and the neglect of its presence could lead to thermodynamical anomalies. Could $\Delta C_p > 0$ in an enthalpy driven reaction leading to a formation of bound state be such an anomaly?

1. Suppose HSP90 has MB can have large C_p and that it is at the temperature of the environment. The temperature varies in the range 2-25 °C being considerably below the physiological temperature 37 C proposed to correspond to a maximal temperature - Hagedorn temperature - for the magnetic flux tubes of SPs. C_p for the MB of SP is expected to increase as the temperature rises since new DFs are thermally excited.

   C_p could be rather high already for the initial state if it corresponds to the sum of heat capacities for nucleotide/ansamycin and HSP90. The size of the MB of nucleotide for $h_{eff} = h$ should be small if it correlates with the size of nucleotide/ansamycin. Nucleotide is an information molecule and therefore its MB should be at a low temperature and have low C_p (thermal energies cannot excite the states at low temperature).

2. Since binding reaction is in question, C_p for the combined system should be reduced unless something happens at the level of MBs. Could the heat capacity of MB of HSP90 increase for nucleotide binding? Could even the value of ΔH for the nucleotide case be larger than thought due to the fact that part of ΔH is transferred to MB of HSP90?

   (a) A lot of heat is liberated in the exothermic binding reaction in both cases. The measure part of the liberated heat goes to the standard DFs discussed in the article. Part of ΔH is transferred to the MB of HSP90 and can heat it to a higher local temperature. New DFs open and heat capacity of MB of CPS90 increases so much that the net heat capacity can increase despite the reduction of ordinary contribution to C_p.

This would happen for the nucleotide but not for ansamycin. Why would the fraction of liberated heat going to the MB of HSP be so small for ansamycin that ΔC_p remains negative?

(b) Could the heat ΔH liberated in the nucleotide case be considerably larger than assumed and larger than for ansamycin plus CSP. This is quite possible since only the fraction going to the environment is measured, not that transferred to MB. Theoretical estimates do not of course take the possible presence MB into account. If ΔH for the nucleotide case is larger than believed, then MB of HSP90 can be heated more and $\Delta C_p > 0$ is possible.

(c) The inspection of tables of [?] shows that the values of ΔH for the nucleotide case are in the range 3-8 eV *per* reaction and correspond to UV energies. For reactions $\Delta C_p < 0$ the values of ΔH are of order .3 eV and correspond to IR photons but with energies larger than thermal energies. The difference is more than order of magnitude and suggests a similar difference for ΔH transferred to MB, which supports the proposed explanation.

2.5 Some functions of SPs in TGD perspective

2.5.1 SPs as heat baths for molecular heat engines and providers of heat energy to ATPs

Heat is produced as a side effect of metabolism and HSPs could extract this heat using remote metabolism and transform it to heat energy resourses liberated when needed.

SPs could be used for heating as in the basic biological processes like transcription and transcription. SPs could also act as heat engines transforming heat energy to work in the case of molecular motors [?].

There are reports about the role of HSPs in doing molecular work [?, ?, ?]: the new element would be heat energy coming from the MB of SP. At least SPs such as HSP60, HSP70, HSP90, HSP104 binding to ATP could serve as general purpose heat engines transforming heat energy at their MB to metabolic energy currency used in various biological processes.

1. All processes produce heat and the very idea of HSPs would the that HSPs gather this heat energy and act as heaters as in the case of transcription, translation, and replication or as heat engines liberating the heat energy as ordered energy. Action as ATPase would make HSP a general purpose molecular heat engine. Currently, we know that HSPX for $X \in \{60, 60, 90, 104\}$ at least act as ATPases.

   Very important question: Is ATPase property a general property of HSPs?*

   By the second law of thermodynamics these heat engines have some maximal efficiency proportional to the difference of the temperatures for heat bath - now MB - and the system receiving the energy. Hence HSP MBs must be at a temperature higher than the systems receiving the energy. The formation of HSP90-ATP bound state would liberate binding energy about 3-7 eV *per* reaction (metabolic energy quantum is .5) eV and this heats MB of HSP and would lead to the reported increase of heat capacity.

2. There is, however, a reason to worry. By Carnot's law maximal effectiveness is proportional to $\Delta T/T$, where ΔT is the temperature difference between the system receiving the work and heat bath, now the MB of SP, and T the temperature of the heat bath. Is the temperature difference high enough to give a reasonable effectiveness?

   ZEO provides a quantum manner to get rid of worries. Time reversal could make possible for the MB of HSP to develop a temperature higher than that of environment by what looks for an observer extraction of thermal energy from the environment but is actually BSFR leading to final state which dissipates in reverse time direction to a state in which the temperatures are equal. T_H should be however somewhat higher than the physiological temperature.

2.5.2 Heat shock protein 70 and ATP in homeostasis

ATP depletes in stress situations due to the lack of ordinary metabolic energy feed as in ischemia. The role of HSP70 and its co-function with ATP in this kind of situation is discussed in [?] . Also HSP70 involves ATPase and the lack of the ordinary metabolic energy could be replaced by thermal metabolic energy feed from the MBs of say HSP70.

2.5.3 SPs and infection

One can distinguish between immune response, which is specific to the invader organism (say bacterium or virus) or molecule and non-specific immune response involving inflammation and fever. Infection includes both the effects of the invader and those caused by the non-specific immune response.

1. The invader specific immune response would be basically an action of the MB: this is the basic vision of TGD. Already the MB of water recognizes the invader molecules by the cyclotron energy spectrum of their MBs: this is just water memory [?, ?, ?] discussed from TGD point view in [?]. "Homeopathy" is he ugly synonym for "water memory" and involves mechanical agitation feeds energy to the MBs of water clusters forming a population mimicking invader molecules.

   MBs of water clusters are varying its flux tube thicknesses and in this manner changing corresponding cyclotron frequencies to get in tune with possible invaders: this is similar to what we do when we search for a radio station. When a hit occurs, MB of the water cluster fixes the flux tube thickness. After getting to resonance, the MBs of water molecules clusters can reconnect with U-shaped flux tubes to corresponding bacterial flux tubes: a pair of flux tubes connecting the water cluster MB to the invader molecule is formed. Invader is caught. The chemical side of the immune system emerged later and would involve sequences of dark proton triplets associated with proteins as addresses - 3N-fold resonance.

2. When bacteria infect cells, they induce inflammation and fever by raising the body temperature as a non-specific immune response. Inflammation can be seen as the body's protective response against infection. The fever helps immune cells to migrate to infection by a process known as chemotaxis. What fever and inflammation could mean in the proposed picture about SPs?

A possible explanation is as follows.

1. Quite generally, the loss of quantum coherence as a reduction of h_{eff} induced by the attack by bacteria should transform ordered energy to heat and produce entropy and also raise the temperature inducing fever. One possible mechanism producing heat in the loss of quantum coherence could be the decay of dark cyclotron condensates and dark photon states to biophotons with $h_{eff} = h$ and with energies around Hagedorn energy of order the energy associated with the physiological temperature. Also the decay of dark proton sequences in the reduction $h_{eff} \to h$ to ordinary protons would liberate energy as photons: Pollack's experiments show that IR irradiation produces exclusion zones (EZs) most effectively so that the energy would be in IR range.

2. Inflammation involves HSPs, in particular HSP70 [?]. If the heat produced by the infection causing the fever can be seen as an entropic waste energy, SPs such as HSP90 would do its best to transform it to ordered energy realized as metabolic energy quanta with the nominal value around .5 eV. As discussed, this would mean a formation of bound states liberating energy - for instance, HMP90-ATP bound state would liberate energy with part going to the MB of HSP70/90 and part to a local environment.

   HMP70/90 acting as ATPases in the bound state and generate metabolic energy quanta by the ADP $\to ATP$ process. The liberated binding energy could cause the observed raise of C_p of the MB of HMP90 and allow it to absorb more effectively heat energy from the environment by temporary time reversal and transform it to metabolic energy quanta. HSPs would be thus generated to absorb the surplus heat to be used as a metabolic energy resource and fever would be reduced as a consequence.

3 Speculative mechanisms explaining some biological observations

In the sequel some speculative applications will be considered.

3.1 Obesity, failing diets, and SPs

The effects of diets on HSP expression and activation have recently been studied, see for instance [?, ?, ?, ?]. During the initial phase of diet the weight is lost. After that the weight often starts to regain. Does a new energy source emerge or is the level of metabolic energy consumption reduced so that the weight regain starts although the nutrient feed stays at the same albeit reduced level?

1. The fractality of TGD Universe suggests an analogy to our society. Living organism is a molecular society, and the fractality of the TGD Universe encourages looking at the situation from the point of view of our own society. Our energy resources have been depleting and we have learned to save energy, and also to recycle thermal energy to increase thermal efficiency. Could the organism learn to use remote metabolism to extract thermal energy from the environment besides SPs. Note that the thermal energy of a thermal photon at room temperature is rather near to the Coulomb energy of a unit charge assignable to the cell membrane voltage perhaps defining another metabolic energy currency.

2. The TGD explanation relies on the proposed ability of at least some SPs to act as ATPases transforming heat energy of their MB to ordered energy realized as metabolic energy quanta. The binding of SPs to ATP [?] would also liberate binding energy transformed to heat, which is partially transferred to MB as heat energy serving as an additional metabolic energy source. Also the reduction of heat losses would mean more effective use of metabolic energy.

   People having obesity predisposition might generate HSP60 or HSP70 and HSP90 even in situations without stress. Also psychological stress such as depression might generate HSPs [?]. HSP60 is known to be associated with obesity [?]. HSP60 is associated with mitochondria and has ATP binding site but does not have ATPase. Could ATP binding site give HSP60 a role analogous to that of ATPase using heat energy of MB of HSP70 to generate ATP from ADP? A more plausible option is that the binding of ATP provides energy for HSP60 and only HSP70 and HSP90 act as ATPases.

   HSP70 [?] expression is considerably higher in obesity without metabolic syndrome but lower in obesity with metabolic syndrome [?]. This would suggest that the diet induces expression of HSP70 and therefore brings an additional metabolic energy source available. In metabolic syndrome the level of HSP70 utilizing thermal energy and reducing entropy of the system would be abnormally low. Besides the expression of HSP70, also its activation is needed [?]. If the activation takes a considerable time, one could understand why it takes time for the additional metabolic energy source to emerge.

3. The ability to act as heat engines and ATPases relies on ZEO: the MBs of SPs could extract thermal energy from the environment in a mode with reversed arrow of time: instead of a disappearance of the necessary temperature gradient it would be generated. One can also say that system learns during diet to use remote metabolism.

   The phenomenon of remote metabolism or quantum credit card has been previously proposed by Pitkänen [?]: system would actively extract energy rather than receive it passively. The receiver of effective negative energy signal would be analogous to a population reversed laser assignable to MB. Quantum credit card would facilitate rapid access to energy via bypassing "bureaucratic formalities". This mechanism applies also to information transfer and makes communications possible with effective signal velocity exceeding the maximal signal velocity.

Quite recently, it has been learned that quite simple physical systems can "breathe" by extracting the energy of Brownian motion [?]: the finding is discussed from the point of view of ZEO in [?].

4. The utilization of metabolic energy becomes more effective during diet and there is less waste of energy. Less nutrients would be required and if the dietary consumption stays at the same albeit reduced levels, fat begins to be regenerated. Dietary stress would induce the generation of SPs. SPs acting as APTases would extract thermal energy from the environment and also from the liberated binding energy in the formation of SP-ATP complex and liberate it as ordered energy by ADP$\to$ATP process. The slow rate for the generation of enzymes needed to generate and activate SPs might be the reason for the slow response

5. One could see the situation also in the following manner. ZEO and time reversal are involved with the extraction of thermal energy from the environment by the MBs of SPs. One can also say that the system learns during the diet to use remote metabolism. The time reversal would be the analog of sleep period. Also we get metabolic energy resources during sleep and the same mechanism could be involved. This could be also seen as hibernation/sleeping at the molecular level and the hibernation/sleep even at the level of organisms could rely on the same mechanism.

3.2 Sleigh dogs which run for days without eating, and starving bacterial colonies

Suppose that the general view about SPs is correct. Assume also there is a fractal hierarchy of MBs. Not only those of biomolecules and of smaller systems, but also of cells, organelles, organs, bodies, larger units like populations...

Assume also that $h_{eff} = h_g r$ holds true so that the cyclotron energy spectrum does not depend on the mass of the dark charged particle. This implies that MBs at all levels of the hierarchy can communicate with the lowest level and also exchange energy and serve as metabolic energy sources. SPs would thus allow the transfer of energy to all these levels.

This admittedly speculative picture could explain the reported ability of sled dogs to run several days without eating [?]: they could store the energy to their MBs and use it during substrate lack. A possible storage to their collective MB would increase further the energy storage ability. This would mean a connection to collective levels of consciousness predicted by TGD and receival of metabolic energy feed as dark photons from these levels [?]. h_{eff} hierarchy indeed makes possible energy transfer and communications between widely different scales characterizing a hierarchy of conscious entities.

This picture could partially explain also why bacteria in media lacking substrate form tightly bound colonies looking like multicellulars. They could store energy to their MB and use it during its substrate lack. Perhaps also the dissipation is reduced because h_{eff} increases.

The cells could also learn to extract thermal energy of the cellular environment besides the thermal energy of SPs, which is more or less another manner to say the same. Starvation could have been the evolutionary pressure leading to the formation of multicellulars. Indeed, the embryos of multicellulars are found to form tightly bound bacterial colonies [?]: the TGD based model is discussed in [?]. There is also anecdotal evidence about analogous abilities of Tibetan monks and people regarded as saints.

To summarize, the proposed general model involves several new physics elements. the new view about space-time and fields, the new view about quantum theory based on ZEO predicting time reversal in BSFRs and a new view about self organization and a realization of SOQC, the h_{eff} hierarchy labelling dark matter as phases of ordinary matter predicted by number theoretic vision about TGD, and the hierarchy of collective levels if consciousness having as a correlate the hierarchy of MBs carrying dark matter in TGD sense. This vision can be defended only by its internal consistency and ability to solve a long list of deep problems of recent day physics.

Received January 19, 2021; Accepted October 1, 2021

References

[Bahrami et al 2019] Bahrami A et al. Depression in adolescent girls: Relationship to serum vitamins a and E, immune response to heat shock protein 27 and systemic inflammation. *J Affect Disord*, 2152:68–73, 2019. doi: 10.1016/j.jad.2019.04.048.

[Biter et al 2013] Biter AB Sielaff B Lee S Tsai Lee J, Kim JH. Heat shock protein (Hsp) 70 is an activator of the Hsp104 motor. *Proc Natl Acad Sci U S A*, 110(21), 2013. doi: 10.1073/pnas.1217988110.

[Bonventre et al 1999] Bonventre JV Mallouk YASM, Vayssier-Taussat MURI and Polla BS. Heat shock protein 70 and ATP as partners in cell homeostasis. *International journal of molecular medicine*, 4(5):463–537, 1999.

[Christopher et al] Christopher Fu S Nilapwar S, Williams E and Ladbury JE. Structural-Thermodynamic Relationships of Interactions in the N-Terminal ATP-Binding Domain. *J. Mol. Biol.*, 392:923–936, 2009. doi:10.1016/j.jmb.2009.07.041.

[Feidantsis et al 2013] Feidantsis K Roufidou C Despoti S Efthimia A, Kentepozidou E and Chatzifotis S. Starvation and re-feeding affect Hsp expression, MAPK activation and antioxidant enzymes activity of European Sea Bass (Dicentrarchus labrax). *Comparative Biochemistry and Physiology Part A: Molecular & Integrative Physiology*, 165(1):79–88, 2013.

[Feidantsis et al 2019] Doyle SM, Hoskins JR, Kravats AN, Heffner AL, Garikapati S, Wickner S. Intermolecular interactions between hsp90 and hsp70. *J Mol Biol*, 431(15):2729–2746, 2019. doi: 10.1016/j.jmb.2019.05.026.

[Fouvet et al 2018] Fauvet B Barducci A De Los Rios P Goloubinoff P, Sassi AS. Chaperones convert the energy from ATP into the nonequilibrium stabilization of native proteins. *P.Nat Chem Biol*, 14(4):388–395, 2018. doi: 10.1038/s41589-018-0013-8.

[Habich et al 2017] Habich C Bouillot JL Eckel J Sell H, Poitou C and Clement K. Heat Shock Protein 60 in Obesity: Effect of Bariatric Surgery and its Relation to Inflammation and Cardiovascular Risk. *Obesity (Silver.Spring.)*, 25(12):2108–2114, 2017.

[Ho 2011] Ho M-W. Quantum Coherent Water Homeopathy, 2011. Available at: http://www.i-sis.org.uk/Quantum_Coherent_Water_Homeopathy.php.

[Kauffman 2015] Kauffman S. Quantum Criticality at the Origin of Life, 2015. Available at: http://arxiv.org/pdf/1502.06880v2.pdf.

[Kellner et al 2019] Kellner R Schuler B De Los Rios P Barducci A Assenza S, Sassi AS. Efficient conversion of chemical energy into mechanical work by Hsp70 chaperones. *Elife*, 8:e48491, 2019. doi: 10.7554/eLife.48491.

[Krause et al 2015] Krause M et al. The chaperone balance hypothesis: The importance of the extracellular to intracellular hsp70 ratio to inflammation-driven type 2 diabetes, the effect of exercise, and the implications for clinical management. *Mediators Inflamm.*, 2015(2015: 249205), 2015. Available at: https://www.ncbi.nlm.nih.gov/pmc/articles/PMC4357135/.

[Kuzkin,Krivtsov 2020] Kuzkin VA and Krivtsov AM. Ballistic resonance and thermalization in the Fermi-Pasta-Ulam-Tsingou chain at finite temperature. *Phys Rev*, 101(042209), 2020. Available at: https://tinyurl.com/y9ycj3nt.

[Mayer et al 2019] Mayer MB, Rosenzweig R, Nillegoda NB and Bukau B. The HSP70 chaperone network. *Nature Reviews: Molecular Cell Biology*, 2019. Available at: https://www.nature.com/articles/s41580-019-0133-3.

[Mills et al 2003] Mills R et al. Spectroscopic and NMR identification of novel hybrid ions in fractional quantum energy states formed by an exothermic reaction of atomic hydrogen with certain catalysts, 2003. Available at: http://www.blacklightpower.com/techpapers.html.

[Montagnier et al 2009] Montagnier L et al. Electromagnetic Signals Are Produced by Aqueous Nanostructures Derived from Bacterial DNA Sequences. *Interdiscip Sci Comput Life Sci* . Available at: http://www.springerlink.com/content/0557v31188m3766x/, 2009.

[Nottale an Da Rocha 2003] Nottale L Da Rocha D. Gravitational Structure Formation in Scale Relativity, 2003. Available at: http://arxiv.org/abs/astro-ph/0310036.

[Nuray and Ferhan 2001] Nuray N and Ferhan E. Cold Shock Proteins. *Turk J Med Sci*, 31:283–290, 2001.

[Petrauskas et al 2011] Petrauskas V Matulis D Toleikis Z, Cimmperman P. Determination of the volume changes induced by ligand binding to heat shock protein 90 using high-pressure denaturation. *Analytical Biochemistry*, 413:171–178, 2011.

[Pitkänen 2010$_a$] Pitkänen M. Physics as Infinite-dimensional Geometry II: Configuration Space Kähler Geometry from Symmetry Principles. *Pre-Space-Time Journal*, 1(4), 2010. See also http://tgdtheory.fi/pdfpool/compl1.pdf.

[Pitkänen 2010$_b$] Pitkänen M. Physics as Infinite-dimensional Geometry I: Identification of the Configuration Space Kähler Function. *Pre-Space-Time Journal*, 1(4), 2010. See also http://tgdtheory.fi/pdfpool/kahler.pdf.

[Pitkänen 2011] Pitkänen M. DNA and Water Memory: Comments on Montagnier Group's Recent Findings. *DNA Decipher Journal*, 1(1), 2011. See also http://tgtheory.fi/public_html/articles/mont.pdf.

[Pitkänen 2013] Pitkänen M. A General Model for Metabolism. *Journal of Consciousness Exploration and Research*, 4(9), 2013. See also http://tgdtheory.fi/public_html/articles/remotetesla.pdf.

[Pitkänen 2014$_a$] Pitkänen M. Music, Biology and Natural Geometry (Part I). *DNA Decipher Journal*, 4(2), 2014. See also http://tgtheory.fi/public_html/articles/harmonytheory.pdf.

[Pitkänen 2014$_b$] Pitkänen M. Music, Biology and Natural Geometry (Part II). *DNA Decipher Journal*, 4(2), 2014. See also http://tgtheory.fi/public_html/articles/harmonytheory.pdf.

[Pitkänen 2014$_c$] Pitkänen M. *Life and Consciousness: TGD based vision*. Lambert. Available at: http://tinyurl.com/zn98vka., 2014.

[Pitkänen 2015$_a$] Pitkänen M. Criticality and Dark Matter. *Pre-Space-Time Journal*, 6(1), 2015. See also https://tinyurl.com/yc3hx4uu, https://tinyurl.com/ybaj3mcj, and https://tinyurl.com/ybkw5qnb.

[Pitkänen 2015$_b$] Pitkänen M. Recent View about Kähler Geometry and Spinor Structure of WCW. *Pre-Space-Time Journal*, 6(4), 2015. See also http://tgtheory.fi/pdfpool/wcwnew.pdf.

[Pitkänen 2015$_c$] Pitkänen M. Hierarchies of Conformal Symmetry Breakings, Quantum Criticalities, Planck Constants, and of Hyper-Finite factors. *Pre-Space-Time Journal*, 6(5), 2015. See also http://tgtheory.fi/public_html/articles/hierarchies.pdf.

[Pitkänen 2016] Pitkänen M. *Topological Geometrodynamics: Revised Edition*. Bentham.Available at: http://tinyurl.com/h26hqul., 2016.

[Pitkänen 2017a] Pitkänen M. Philosophy of Adelic Physics. Available at: http://tgdtheory.fi/public_html/articles/adelephysics.pdf., 2017.

[Pitkänen 2017b] Pitkänen M. On Hydrinos Again. *Pre-Space-Time Journal*, 8(1), 2017. See also http://tgtheory.fi/public_html/articles/Millsagain.pdf.

[Pitkänen 2018] Pitkänen M. On the Physical Interpretation of the Velocity Parameter in the Formula for Gravitational Planck Constant. *Pre-Space-Time Journal*, 9(7), 2018. See also http://tgtheory.fi/public_html/articles/vzero.pdf.

[Pitkänen 2019a] Pitkänen M. Self-organization by h_{eff} Changing Phase Transitions. *Pre-Space-Time Journal*, 10(7), 2019. See also http://tgtheory.fi/public_html/articles/heffselforg.pdf.

[Pitkänen 2019b] Pitkänen M. Scattering Amplitudes & Orbits of Cognitive Representations under Subgroup of Symplectic Group Respecting the Extension of Rationall. *Pre-Space-Time Journal*, 10(4), 2019. See also http://tgtheory.fi/public_html/articles/symplorbsm.pdf.

[Pitkänen 2020a] Pitkänen M. Summary of Topological Geometrodynamics. Research Gate: https://www.researchgate.net/publication/343601105_Summary_of_Topological_Geometrodynamics., 2020.

[Pitkänen 2020b] Pitkänen M. Multilocal viruses. *DNA Decipher Journal*, 10(1), 2020. See also http://tgtheory.fi/public_html/articles/viralstrange.pdf.

[Pitkänen 2020c] Pitkänen M. Zero Energy Ontology & Consciousness. *Journal of Consciousness Exploration & Research*, 11(1), 2020. See also http://tgtheory.fi/public_html/articles/zeoquestions.pdf.

[Pitkänen 2020d] Pitkänen M. Ballistic resonance and zero energy ontology. Research Gate: https://www.researchgate.net/publication/343601103_Ballistic_resonance_and_zero_energy_ontology., 2020.

[Robson 2008] Robson D. Researchers Seek to Demystify the Metabolic Magic of Sled Dogs, 2008. Available at: http://tinyurl.com/o4o8srm.

[Renes et al 2016] Renes J Bouwman FG Westerterp KR Roumans N. J., Camps SG and Mariman EC. Weight loss-induced stress in subcutaneous adipose tissue is related to weight regain. *Br J Nutr*, 115(5):913–920, 2016.

[Saad, Sabbah and Rezk 2019] Saad MSS Sabbah NA, Rezk NA. Relationship Between Heat Shock Protein Expression and Obesity With and Without Metabolic Syndrome. *Genet Test Mol Biomarkers*, 23(10):737–743, 2019. Available at: https://pubmed.ncbi.nlm.nih.gov/31517511/.

[Smith 2001] Smith C. *Learning From Water, A Possible Quantum Computing Medium*. CHAOS, 2001.

[Szaz et al 2003] Szaz O Szasz A, Vincze G and Szasz N. An energy analysis of extracellular hyperthermia. *Electromagnetic biology and medicine*, 22(2&3):103–115, 2003.

[Wang et al 2017] Wang LC et al. Highly Selective Activation of Heat Shock Protein 70 by Allosteric Regulation Provides an Insight into Efficient Neuroinflammation Inhibition. *EBioMedicine*, 23:160–172, 2017. doi: 10.1016/j.ebiom.2017.08.011.

[Yin et al 2019] Yin et al. The Early Ediacaran Caveasphaera Foreshadows the Evolutionary Origin of Animal-like Embryology. *Current Biology*, 2019. doi: https://doi.org/10.1016/j.cub.2019.10.057. See also http://tinyurl.com/qkzwk5t.

[Zeeman 1977] Zeeman EC. *Catastrophe Theory*. Addison-Wessley Publishing Company, 1977.

[Wikipedia$_a$] Cold shock domain. Available at: https://en.wikipedia.org/wiki/Cold_shock_domain.

[Wikipedia$_b$] Cold shock response. Available at: https://en.wikipedia.org/wiki/Cold_shock_response.

[Wikipedia$_c$] Heat-shock protein. Available at: https://en.wikipedia.org/wiki/Heat_shock_protein.

[Wikipedia$_d$] HSP60. Available at: https://en.wikipedia.org/wiki/HSP60.

[Wikipedia$_e$] HSP60. Available at: https://en.wikipedia.org/wiki/HSP70.

[Wikipedia$_f$] HSP60. Available at: https://en.wikipedia.org/wiki/HSP90.

[Wikipedia$_g$] Molecular motor. Available at: https://en.wikipedia.org/wiki/Molecular_motor.

TGD View about Language

M. Pitkänen[1] and R. Rastmanesh[2,3]

[1]Independent researcher. [1]
[2]Member of The Nutrition Society, London, UK.
[3]Member of The American Physical Society, USA.

Abstract

Human languages differ dramatically from their analogos for animals. Animal languages consist mainly of simple signals, warnings and threats for instance. The emotional expression dominates. There seems to be no grammar. Birds can have repertoire of different song patterns and monkeys have gesture language. There is a huge variety of human languages. One can also regard music as a kind language expressing emotions and creating them. Also pictures define linguistic representations. Children and animals learn speech by mimicry and the grammar and syntax without conscious efforts. Human language is also special in that it involves conceptualization, metaphors, and analogies representing abstract concepts in terms of objects and actions of the external world. One might understand the semantic aspect of language in terms of association and conditioning. Language acquisition involves showing the object and saying the word describing it. This suggests conditioning and association so that a mere word generates an imagined percept of the object. Conditioning and formation of associations is a very general form of learning assumed to relate to the increase of synaptic strengths leading to a generation of association pathways. In computer science pattern recognition and completion models it mathematically.

Amazingly, only a few point mutations for relatively few genes seem so have led to human languages and transformed biological evolution to cultural evolution? What happened for these genes? In the biochemistry framework it is difficult to imagine an answer to this question. Here TGD could come in rescue. Number theoretic physics is part of quantum TGD and essential for understanding evolution as an increase of algebraic complexity. Evolutionary hierarchies would correspond to hierarchies of algebraic extensions of rationals. The dimension n of extension defines effective Planck constant $h_{eff}/h_0 = n$. The larger the dimension, the larger the scale of quantum coherence at the corresponding layer of the magnetic body (MB) associated with the system: n would be analogous to IQ. One can assign a value of h_{eff} characterizing the evolutionary level also to genes. The genes with larger h_{eff} would serve as control genes and the increase of h_{eff} would mean an evolutionary step. Perhaps a dramatic increase of h_{eff} occurred to FOXP2 and some other genes as human language emerged.

1 Introduction

Human languages differ dramatically from their analogos for animals. Animal languages consist mainly of simple signals, warnings and threats for instance. The emotional expression dominates. There seems to be no grammar and syntax unlike in human languages. Birds can have impressive repertoire of different song patterns and monkeys have gesture language.

There is a huge variety of human languages: speech and written language, sign languages based on gestures, the language of mathematics and computer languages in which emotional expression is absent. One can also regard music as a kind language expressing emotions and creating them. Also pictures define linguistic representations. Children and animals learn language by mimicry and also learn the grammar and syntax without conscious efforts. Adults can learn a foreign language by learning the vocabulary and

---

[1]Correspondence: Matti Pitkänen http://tgdtheory.com/. Address: Rinnekatu 2-4 A8, 03620, Karkkila, Finland. Email: matpitka6@gmail.com. Email: matpitka6@gmail.com.

the rules of grammar. Human language is also special in that it involves conceptualization, metaphors, and analogies representing abstract concepts in terms of objects and actions of the external world.

One might understand the semantic aspect of language in terms of association and conditioning. Language acquisition involves showing the object and saying the word describing it. This suggests that conditioning and association happens so that mere word generates an imagined percept of the object. Conditioning and formation of associations is a very general form of learning assumed to relate to the increase of synaptic strengths leading to a generation of association pathways. In computer science pattern recognition and completion models it mathematically. One one can ask whether the learning of language and language understanding is something more than this.

For more detailed approaches of language theories, interested readers may be referred to references [11, 13, 12, 16]. The article of Kempe and Brooks [15] and the review article "From Molecule to Metaphor: A neural theory of language" about the language theory of Jerome A. Feldman by Stefan Frank [14] gives a deeper perspective to language theories. The notion of embodiment is in key role in these theories and will be in a key role also in the proposal to be discussed.

1.1 About language genes

Forkhead box protein P2 (FOXP2) encodes a transcription factor involved in language acquisition and speech [7]. In addition to FOXP2 a limited number of genes are involved in speaking [9]. All vertebrates possess FOXP2, however it is estimated that some 120,000-200,000 thousand years ago, some mutations occurred only in humans which aided humans to start initial forms of speaking [11]. Animals have their own primitive language; both voices and gestures with meaning make communications possible. They mainly recognize each other and communicate with pheromones. As for vocabulary, a short review of the Old Testament, cuneiform writings, glossary of old books, and hieroglyphs clearly shows that the number of entries was quite limited in the past. Therefore, a further progression of language could be a matter of cultural communications and technological advances.

However, today it is clear that crucial mutations occurred in the non-coding part of the genome controlling the expression of genes coding for proteins [9] which lead to language evolution. Therefore, the evolutionary step was associated with control of existing genes. Humans are also distinguished from animals by their learning abilities.

Language acquisition must rely on conditioning/associations between language expressions and experiences. It seems that embodiment is the mechanism, which associates to a linguistic expression and imagined sensory percept and/or motor action making the emergence of meaning. What is needed is long term memory and also some kind of standardization of percepts so that they consist of standardized mental images. Pattern recognition and completion could give this standardization.

Since sensory and motor imagination could be seen as almost sensory experiences and almost motor actions, this suggests that new communications between auditory organs and sensory and motor areas emerged. Even more generally, this kind of communication could have emerged quite generally. This would be essentially a new form of conditioning and the same mechanism could apply to all kinds of conditionings.

1.2 How the mutation of only a few genes led to cultural evolution?

Amazingly, only a few mutations for relatively few genes seems so have led to human languages. Why few point mutations of relatively few genes could have transformed biological evolution to cultural evolution? What happened for these genes? In the biochemistry framework it is difficult to imagine an answer to this question. Here TGD could come in rescue.

Number theoretic physics is part of quantum TGD and essential for understanding evolution as an increase of algebraic complexity. Evolutionary hierarchies would correspond to hierarchies of algebraic extensions of rationals. The dimension n of extension defines effective Planck constant $h_{eff}/h_0 = n$, the larger the dimension, the larger the scale of quantum coherence at corresponding level of magnetic body

(MB) associated with the system. One can also say that n is analog of IQ. One can assign a value of h_{eff} characterizing their evolutionary level also to genes. The genes with larger h_{eff} would serve as control genes. The increase of h_{eff} for genes would mean an evolutionary step. Perhaps a dramatic increase of h_{eff} occurred to FOXP2 and some other genes as human language emerged.

Second mechanism could be energy resonance in the coupling of the analogs of DNA, RNA, tRNA, and amino acids consisting of dark proton triplet with their chemical counterparts. The coupling would be between the entire gene and its dark analog and codon sequence would play a role of address. In both cases small changes of the gene could spoil or produce an energy resonance. This sensitivity would make genes an ideal control tool but would also serve as a general mechanism also for genetic diseases. The increase of h_{eff} accompanied by a small mutation to guarantee energy resonance could be the mechanism explaining the importance of FOXP2 and similar control genes.

2 Number theoretical aspects of quantum biology

The basic ideas about consciousness and life are discussed in Appendix. Here the aspects relevant for the recent work are discussed.

2.1 Dark proton representation of genetic code

The model for codons of genetic code emerged from the attempts to understand water memory [18]. The outcome was a totally unexpected finding [22, 18]: the states of dark nucleons formed from three quarks connected by color bonds can be naturally grouped to multiplets in one-one correspondence with 64 DNAs, 64 RNAs, 20 amino acids, and tRNA and there is natural mapping of DNA and RNA type states to amino acid type states such that the numbers of DNAs/RNAs mapped to given amino acid are same as for the vertebrate genetic code.

The basic idea is simple. The basic difference from the model of free nucleon is that the nucleons in question - maybe also nuclear nucleons - consist of 3 linearly ordered quarks - just as DNA codons consist of three nucleotides. One might therefore ask whether codons could correspond to dark nucleons obtained as open strings with 3 quarks connected by two color flux tubes or as closed triangles connected by 3 color flux tubes. Only the first option works without additional assumptions. The codons in turn would be connected by color flux tubes having quantum numbers of pion or η.

This representation of the genetic would be based on entanglement rather than letter sequences. Could dark nucleons constructed as a string of 3 quarks using color flux tubes realize 64 DNA codons? Could 20 amino acids be identified as equivalence classes of some equivalence relation between 64 fundamental codons in a natural manner? The codons would not be separable to letters but entangled states of 3 quarks anymore.

If this picture is correct, genetic code would be realized already at the level of dark nuclear physics and maybe even in ordinary nuclear physics if the nucleons of ordinary nuclear physics are linear nucleons. Chemical realization of genetic code would be induced from the fundamental realization in terms of dark nucleon sequences and vertebrate code would be the most perfect one. Chemistry would be a kind of shadow of the dynamics of positively charged dark nucleon strings accompanying the DNA strands and this could explain the stability of the DNA strand having 2 units of negative charge per nucleotide. Biochemistry might be controlled by the dark matter at flux tubes.

The ability of the TGD based model to explain genetic code in terms of spin pairing is an impressive achievement still difficult to take quite seriously - perhaps it is better so!

1. The original model mapping codons to dark nucleon states assumed the overall charge neutrality of the dark proton strings: the idea was that the charges of color bonds cancel the total charge of dark nucleon so that all states uuu, uud, udd, ddd can be considered. The charge itself would not affect the representation of codons. Neutrality assumption is however not necessary. The interpretation as

dark nucleus resulting from dark proton string could quite well lead to the formation of the analog of ordinary nucleus via dark beta decay [29] so that the dark nucleus could have charge. Isospin symmetry breaking is assumed so that neither quarks nor flux tubes are assigned to representations of strong $SU(2)$.

There is a possible objection. For ordinary baryons the mass of Δ is much larger than that of the proton. The mass splitting could be however much smaller for linear baryons if the mass scale of excitations scales as $1/h_{eff}$ as indeed assumed in the model of dark nuclear strings [28, 29].

2. The model assumes that the states of DNA can be described as tensor products of the four 3-quark states with spin content $2 \otimes 2 \otimes 2 = 4 \oplus 2_1 \oplus 2_2$ with the states formed with the 3 spin triplet states $3 \otimes 3 = 5 \oplus 3 \oplus 1$ with *singlet state dropped*. The means that flux tubes are spin 1 objects and only spin 2 and spin 1 objects are accepted in the tensor product. One could consider interpretation in terms of ρ meson type bonding or gluon type bonding. With these assumptions the tensor product $(2 \otimes 2 \otimes 2) \otimes (5 \oplus 3)$ contains $8 \times 8 = 64$ states identified as analogs of DNA codons.

The rejection of spin 0 pionic bonds looks strange. These would however occur as bonds connecting dark codons and could correspond to different p-adic length scale as suggested by the successful model of X boson [30].

One can also ask why not identify the dark nucleon as as a closed triangle so that there would be 3 color bonds. In this case $3 \otimes 3 \otimes 3$ would give 27 states instead of 8 ($\oplus 1$). This option does not look promising.

3. The model assumes that amino acids correspond to the states 4×5 with $4 \in \{4 \oplus 2 \oplus 2\}$ and $5 \in \{5 \oplus 3\}$. One could tensor product of spin 3/2 quark states and spin 2 flux tube states giving 20 states, the number of amino acids!

4. Genetic code would be defined by projecting DNA codons with the same total quark and color bond spin projections to the amino acid with the same (or opposite) spin projections. The attractive force between parallel vortices rotating in opposite directions serves as a metaphor for the idea. This hypothesis allows immediately the calculation of the degeneracies of various spin states. The code projects the states in $(4 \oplus 2 \oplus 2) \otimes (5 \oplus 3)$ to the states of 4×5 with the same or opposite spin projection. This would give the degeneracies $D(k)$ as products of numbers $D_B \in \{1, 2, 3, 2\}$ and $D_b \in \{1, 2, 2, 1\}$: $D = D_B \times D_b$. Only the observed degeneracies $D = 1, 2, 3, 4, 6$ are predicted. The numbers $N(k)$ of amino acids coded by D codons would be

$$[N(1), N(2), N(3), N(4), N(6)] = [2, 7, 2, 6, 3] \ .$$

The correct numbers for vertebrate nuclear code are $(N(1), N(2), N(3), N(4), N(6)) = (2, 9, 1, 5, 3)$. Some kind of symmetry breaking must take place and should relate to the emergence of stopping codons. If one codon in the second 3-plet becomes stopping codon, the 3-plet becomes doublet. If 2 codons in 4-plet become stopping codons it also becomes doublet and one obtains the correct result $(2, 9, 1, 5, 3)$!

It is difficult to exaggerate the importance of this simple observation suggesting that genetic code is realized already at the level of dark or even ordinary nuclear physics and bio-chemistry is only a kind of shadow of dark matter physics.

2.2 Bio-harmony as a realization of genetic code

TGD leads to a notion of bio-harmony in terms of icosahedral and tetrahedral geometries and 3-chords made of light assigned to the triangular faces of icosahedron and tetrahedron [26, 27, 45]. The surprise was that vertebrate genetic code emerged as a prediction: the numbers of DNA codons coding for a given amino acid are predicted correctly. DNA codons correspond to triangular faces and the orbit of a given triangle under the symmetries of the bio-harmony in question corresponds to DNA codons coding for the amino acid assigned with the orbit.

Codon corresponds to 6 bits: this is information in the usual computational sense. bio-harmony codes for mood: emotional information related to emotional intelligence as ability to get to the same mood allowing to receive this information. bio-harmony would be a fundamental representation of information realized already at molecular level and speech, hearing and other expressions of information would be based on it. For emotional expression at RNA level possibly involved with conditioning at synaptic level see [35].

Does the generation of nerve pulse patterns by a gene mean at the cell membrane from dark DNA to dark protein map to dark protein (it could be also dark RNA or dark DNA even) associated with the cell membrane. What about communications with RNA and enzymes involved with transcription and translation. Do all basic biocatalytic processes involve them.

What about a generalization of Josephson currents. Dark ions certainly define them but could also dark proton triplets and their sequences associated with proteins give rise to oscillating Josephson currents through cell membrane and therefore to dark Josephson radiation with 3N dark photon units! Proteins themselves need not move much!

The universal language could be restricted to the genetic code which would be realized by dark proton triplets. The 64 codons are formed from 3 20-chord harmonies associated with icosahedron and the unique 4-chord harmony associated with tetrahedron. Bio-harmonies are associated with the so-called Hamiltonian cycles ,which go through every vertex of Platonic solid once. For icosahedron the number of vertices is 12, the number of notes in 12-note scale.

Also tetrahedron, cube, octahedron and dodecahedron are possible and one can consider the possibility that they also define harmonies in terms of Hamiltonian cycles. Dodecahedron would have 5-chords (pentagons as faces) as basic chords and there is only single harmony. Same mood always, very eastern and enlightened as also the fact that scale would have 20 notes.

Also octahedron gives 3-chords (triangular faces) whereas cube gives 4-chords (squares as faces). One can of course speculate with the idea that DNA could also represent this kind of harmonies: sometimes the 3N rule is indeed broken, for instance for introns.

Galois confinement allows the possibility of interpreted dark genes as sequences of N dark proton triplets as higher level structures behaving like a single quantal unit. This would be true also for the corresponding dark photon sequences consisting of 3N dark photons representing the gene in bio-harmony as an analog of a music piece consisting of 3-chords and played by transcribing it to mRNA.

One can be even more general. Any discrete structure, defining graph, in particular cognitive representation providing a unique finite discretization of space-time surface as points with the coordinates of the 8-D imbedding space coordinates in the extension of rationals, defines harmonies in terms of Hamiltonian cycles. Could also these harmonies make sense? The restrictions of the cognitive representations to 2-D partonic 2-surfaces would define something analogous to bio-harmony as Hamiltonian cycle of 2-D graph (Platonic surfaces solids can be regarded as 2-D graphs). Also now the interpretation as representations of Galois groups and the notion of Galois confinement would be important.

2.2.1 About the details of the genetic code based on bio-harmony

TGD suggests several realizations of music harmonies in terms of Hamiltonian cycles representing the notes of music scale, most naturally 12-note scale represented as vertices of the graph used. The most plausible realization of the harmony is as icosahedral harmony [26, 27] (see http://tinyurl.com/yad4tqwl and http://tinyurl.com/yyjpm25r).

1. Icosahedron (see http://tinyurl.com/l5sphzz) has 12 vertices and Hamiltonian cycle as a representation of 12-note scale would go through all vertices such that two nearest vertices along the

cycle would differ by quint (frequency scaling by factor 3/2 modulo octave equivalence). Icosahedron allows a large number of inequivalent Hamiltonian cycles and thus harmonies characterized by the subgroup of the icosahedral group leaving the cycle invariant. This group can be Z_6, Z_4, or Z_2 which acts either as a reflection group or corresponds to a rotation by π.

2. The fusion of 3 icosahedral harmonies with symmetry groups Z_6, Z_4 and Z_2 gives 20+20+20=60 3-chords and $3+1+5+10$ =19 orbits of these under symmetry group and almost vertebrate genetic code when 3-chords are identified as analogs of DNA codons and their orbits as amino acids. One obtains counterparts of 60 DNA codons and $3+1+5+10$ =19 amino acids so that 4 DNA codons and 1 amino acid are missing.

3. The problem disappears if one adds tetrahedral harmony with 4 codons as faces of tetrahedron and 1 amino acid as the orbit of the face of tetrahedron. One obtains 64 analogs of DNA codons and 20 analogs of amino acids: this harmony was coined as bio-harmony in [26, 27]. The predicted number of DNA codons coding for given amino acid is the number of triangles at the orbit of a given triangle and the numbers are those for genetic code.

4. How to realize the fusion of harmonies? Perhaps the simplest realization found hitherto is based on the union of a tetrahedron of 3 icosahedrons obtained by gluing tetrahedron to icosahedron along its face which is a triangle. The precise geometric interpretation of this realization has been however missing and several variants have been considered. The model could explain the two additional amino acids Pyl and Sec appearing in Nature [26, 27] as being related to different variant for the chemical counterparts of the bio-harmony.

There is also a slight breaking of symmetries: ile 4-plet breaks into ile triplet and met singlet and trp double breaks into stop and trp also leu 4-plet can break in leu triplet and ser singlet (see http://tinyurl.com/puw82x8). This symmetry breaking should be understood.

2.2.2 Cell membrane and microtubules as a higher level representation of genetic code?

Also the realization of genetic code at the level of cell membrane can be considered [36]. The motivation for the current proposal is that the lipids have at their ends negatively charged phosphates just as DNA nucleotides have. The generalization of DNA as a 1-D lattice like structure to a 2-D cylindrical lattice containing nucleotide like units - letters - possibly assignable to lipids and realized as dark protons. Single lipid could be in the role of ribose+nucleotide unit and accompanied by a neutralizing and stabilizing dark proton. For axons one would have cylindrical lattice dark DNA lattice. The two lipid layers could correspond to two DNA strands: the analogs of the passive and active strand.

The finding is that membrane affects protein's behavior. This would be understandable in the proposed pictures 2-D analog of 1-D nucleotides sequences with codons replaced with counterparts of genes as basic units. That lipids are accompanied by phosphates with charge -1 gives the hint. Phosphate charge is neutralized by a dark proton as an analog of a nucleotide. The proposed model for language requires that genes are realized at the lipid layers and the simplest assumption is that genes defined by basic units as sequences of dark protons at the surface covering it.

Genes assigned to the lipid layers of the cell membrane could communicate using dark 3N-photon sequences with the proteins, genome, RNA and DNA. Dark control genes could initiate a nerve pulse pattern. An interesting possibility is that ganglions, nucleus like structures assignable to sensory organs and appearing as basal ganglia in brain [3] (https://cutt.ly/zfWoBFt) could communicate with genes.

Also microtubules have GTPs with charge -3 bound to tubulins. In dynamical instability known as treadmilling the transformation of GTP$\rightarrow$ GDP bound to β tubulin by hydrolysis induces the shortening of the microtubule at minus end whereas the addition of tubulins bound to GTP induces the growth at plus end. Also actin molecules bound to ATP show a similar behavior. Could they be accompanied by dark DNA codons? Are all codons allowed or does the absence of XTP, X= T,C,G mean that only codons of type GGG would be present?

For the dark codons for the cell membrane the p-adic length scale $L(151) \simeq 10^{-8}$ m would correspond to the lipid's transversal size scale and would be the distance between the dark protons. The scale of dark nuclear energy would be proportional to $1/L(151)$ and scaled down by factor $\sim 10^{-3}$ from that for DNA. The energy scale should be above the thermal energy at room temperature about .025 eV. If the energy scale is 2.5 eV (energy of visible photon) for DNA, the condition is satisfied. Note that 2.5 eV is in the bio-photon energy range. For p-adic large scales longer than $L(151)$ thermal instability becomes a problem.

It is interesting to compare the number of codons per unit length for ordinary genetic code (and its dark variant) and for various membranes and microtubules.

- For the ordinary genetic code there are 10 codons per 10 nm defining p-adic length scale $L(151)$. This gives a codon density $dn/dl = 10^3/\mu m$ in absence of coiling. The total number of codons in human DNA with a total length $L \sim 1$ meter is of order $N \sim 10^9$ codons. The packing fraction of DNA due to coiling is therefore huge: of order 10^6.

- If each lipid phosphate is accompanied by a dark proton and if lipid correspond to square at axonal cylinder with side of length $d = L(151)$ and the radius R of axon corresponds to the p-adic length scale $L(167) = 2.5\mu$ m (also of the same order as nucleus size), there are about $dn/dl = 2\pi (R/d)^2 \sim (2\pi/3) \times 10^4 \sim 1.3 \times 10^5/\mu m$. Axon should have length $L \sim 1$ cm to contain the entire genome.

  The same rough estimate applies to microtubules except that there would be one codon per GTP so that the estimate would be 3 times higher if GTP corresponds to length scale $L(151)$ of tubulin molecule. It has been proposed that genetic code is realized at the microtubular level.

- The nuclear membrane assumed to have a radius about $L(167) = 2.5\mu$m could represent $N \sim (4/3)R^2/d^2 \sim .8 \times 10^5$ codons. This is a fraction 10^{-5} about the total number of codons. For a neuronal membrane with radius $R \sim 10^{-4}$ meters assignable to a large neuron the fraction would be roughly 10^{-1}. The fraction of dark codons associated with membranes could correspond to genes involved with the control and communication with genome and other cell membranes. Note that the non-coding intronic portion dominates in the genome of higher vertebrates. One can ask whether the chromosome structure is somehow visible in the membrane genome and microtubular genome.

2.3 Galois group of space-time surface as new discrete degrees of freedom

2.3.1 Galois confinemenent

The problem is to understand how dark photon triplets occur as asymptotic states - one would expect many-photon states with a single photon as a basic unit. The explanation would be completely analogous to that for the appearance of 3-quark states as asymptotic states in hadron physics - the analog of color confinement [50]. Dark photons would form Z_3 triplets under the Z_3 subgroup of the Galois group associated with corresponding space-time surface, and only Z_3 singlets realized as 3-photon states would be possible.

The invariance under $Gal(F)$ would correspond to a special case of Galois confinement, a notion introduced in [49] with physical motivations coming partially from the TGD based model of genetic code based on dark photon triplets.

2.3.2 Cognitive measurement cascades

Quantum states form Galois group algebra - wave functions in Galois group of extension E. E has in general decomposition of extension E_1 as extension of E_2 as extension of ... to a series . Galois group of E has decomposition to product of $Gal(E) = Gal(E/E_1)Gal(E_1)$ and same decomposition holds true for $Gal(E_1)$ so that one has hierarchy of normal subgroups corresponding extension of extension of...hierarchy defined by a composite polynomial $P(x) == P_1(P_2(x))$ with P_2 having similar representation. P defines in M^8 picture the space-time surface. This maps a tensor product composition for group algebra and the factors of group algebra entangle. SSFR corresponds to a quantum measurement cascade: SSFR in $Gal(E/E_1)$, SSFR in $Gal(E_1/E_2)$ etc.

Could this cascade relate to the parsing of a linguistic expression? It would certainly correspond to a sentence S_1 about a sentence S_2 about ... such that one substitutes a concrete sentence for S_1 first, then to S_2, etc.... The sentences in the sequence indeed have h_{eff} which decreases. This is the case in the cascade of SSFRs since $h_{eff}/h_0 = n$ is the dimension of E_n.

I also mentioned the number theoretic measurement cascades for purely number theoretic Galois degrees of freedom. http://tgdtheory.fi/public_html/articles/SSFRGalois.pdf.

Could cascade of flux tubes decaying to smaller flux tubes with smaller value of h_{eff} should correspond to this hierarchy. Certainly this is linguistics but the sentence as argument could correspond to several sub-sentences - different flux tubes. Could a neural pathway defined by the branching axon correspond to a concretization of this kind statement about statement (or multistatement, perhaps nerve pulse pattern generated by nerve pulse patterns arriving to a given neuron) about...

2.4 Energy and frequency resonance as basic elements of dark photon communications

Dark photon realization of genetic code leads to a view about fundamental linguistic communication based on resonance and we will write a separate paper connecting TGD with language soon. Two systems can be in communication when there is resonance. $E = h_{eff} f$ and energy conservation implies

$$h_{eff,1} f_1 = h_{eff,2} f_2 .$$

For $h_{eff,1} = h_{eff,2}$, energy conservation implies that both energies and frequencies are identical: $E_1 = E_2$ and $f_2 = f_2$. Both energy and frequency resonances in question.

In the general case one has $f_1/f_2 = h_{eff,2}/h_{eff,1}$ and frequency scaling takes place. The studies of water memory lead to the observation that this kind of phenomenon indeed occurs [1]. The communications of dark matter with ordinary matter and those between different values of h_{eff} involve only energy resonance. Frequency and wavelength scaling makes it possible for long scales to control short scales. Dark photons with EEG frequencies associated with the big part of MB transform to bio photons with a wavelength of say cell size scale and control dynamics in these short scales: for instance, induce molecular transitions. This is impossible in standard physics.

The resonance condition becomes even stronger if it is required there is a large number of biomolecules in resonance with dark matter realized as dark variants of biomolecules and dark ions. Cyclotron resonance energies are proportional to $\hbar_{eff}$ characterizing magnetic flux tubes and to the valued of the magnetic field strength dictated by the quantization of the monopole flux quantization by the thickness of the flux tube which can be do some degree varied by varying the thickness of the flux tube giving rise to frequency modulation.

The findings of Blackman et al [6] suggest that $B_{end} = 0.2$ Gauss defines an important value in the spectrum of B_{end} values. It could correspond to the field strength for the monopole flux part of the Earth's magnetic field: besides this there would be a non-monopole flux part allowed also in the Maxwellian theory.

There are however indications that the value B_{end} is quantized and is proportional to the inverse of a biologically important p-adic length scale and thus would be quantized in octaves. This could relate directly to the octave equivalence phenomenon in music experience. The model of bio-harmony [26, 27, 45] suggests a further quantization of the octave to Pythagorean 12-note scale of music. This would not be only essential for the music experience but communications of emotions and molecular level using the music of light.

2.4.1 Selection of basic biomolecules by energy resonance

The dark particles must have energy resonance with bio-molecules in order to induce their transitions. This seems to pose extremely strong conditions possibly selecting the bio-molecules able to form interacting networks with dark matter and with each other. One expects that only some amino acids and DNA type molecules survive.

Nottale's hypothesis provides a partial solution to these conditions. Nottale proposed the notion of gravitational Planck constant

$$\hbar_{gr} = \frac{GMm}{v_0}$$

assignable in TGD to gravitational flux tubes connecting large mass M and small mass m and v_0 is velocity parameter. The gravitational flux tube presumably carries no monopole flux. The TGD based additional hypothesis that one has equals to

$$hbar_{gr} = h_{eff} = nh_0 \ .$$

This implies that the cyclotron energy spectrum

$$E_c = n\hbar_{gr}\frac{eB}{m} = n\frac{GM}{v_0}eB$$

of the charged particle does not depend at all on its m. Therefore in a given magnetic field, say B_{end}, the cyclotron resonance spectrum is independent of the particle.

The energy resonance condition reduces to the condition that the charged ion or molecule has some cyclotron energy coming as a multiple of fundamental in its spectrum in the spectrum of its transition energies. Even this condition is very strong since the energy scale for cyclotron energy in B_{end} is in the bio-photon energy range containing energies in visible and UV. The fact that bio-photons have a quasi-continuous spectrum strongly suggests that B_{end} has a spectrum. The model of bio-harmony [26, 27, 45] suggests that the values of B_{end} correspond to Pythagorean scaling constructible by quint cycle.

The above simplified picture is formulated for single dark photon communications. The dark proton and dark photon realizations of the genetic code requires 3-resonance that is a simultaneous energy resonance for the 3 members of dark photon triplet. In dark-dark pairing also frequency resonance is possible. In dark-ordinary pairing frequency increases and couples long scales with short scales. Also resonant communications between genes with N codons involving $3N$ dark photon frequencies must be possible. This requires new physics provided by number theoretical vision.

2.4.2 What happens in the cyclotron resonance?

3 cyclotron energies for flux tubes characterize dark 3-proton triplet and Nottale's hypothesis predicts that they depend on the values of B_{end} for the flux tubes only. bio-harmony suggests that the spectrum of frequencies and thus B_{end} corresponds to Pythagorean 12-note scale for a given octave. The allowed chords of bioharmy would characterize the emotional state at the molecular level and correspond to the holistic emotional aspects of the communication beside the binary information.

The resonance would require that the dark cyclotron energy changes are equal to corresponding energies in molecular transitions. Galois confinement [49] makes possible also 3-N resonance. The resonance condition would select basic biomolecules and the ability of dark analogs of biomolecules to simultaneously resonate with several biomolecules would give additional conditions. In particular this would select DNAs and amino acids.

An open question is whether the coupling to ordinary biomolecules involves a transformation of a dark photon triplet or an N-plet to a single ordinary photon. For instance, does the sum of the 3 cyclotron excitation energies appear in the coupling of dark 3-proton state to amino acid in protein? This would have an analog as 4-wave coupling in laser physics allowing in biology the transformation of dark photon triplet to single biophoton/or 3 bio-photons or vice versa. 6-wave coupling of laser physics would be analogous to the coupling of ordinary 3-photon state to dark 3-photon and back to ordinary 3-photon state.

The resonance itself would mean a process in which dark 3-proton cyclotron excitation returns to the ground state and generates dark 3-photon transforming transforming to ordinary photon (or 3-photon) and absorbed by the ordinary codon or amino acid excitation to hither energy state. This state would in turn emit an ordinary photon transforming to dark 3-photon absorbed by dark codon. This mechanism generalizes to 3N-proton states representing genes or dark proteins.

3 TGD based view about brain

3.1 A new view about the role of nerve pulses in sensory perception

Sensory perception would in TGD generate sensory mental images at sensory organs: this would solve a basic problem of neuroscience due to the similarity of neural tissue in various sensory areas. The new view about time and memory implied by ZEO solves the problem caused by the phantom limb. The pain in the phantom limb is a sensory memory of pain.

The stimulation of temporal lobes indeed generates sensory memories, and people with a cognitive impairment are known for memory feats such as being able to draw a building seen in the past with every detail or to learn music pieces with single listening. These feats can be understood if the memories correspond to "seeing" in time direction with a beam of dark photons travelling to the past reflected back. ZEO allows this.

Since perception involves a lot of processing this would require forth-and back signaling between brain and sensory organs. There would be virtual sensory input from the brain or via the brain. Sensory percept would be an artwork, standardized mental image, resulting as pattern recognition assigning to sensory input standardized mental image nearest to the input.

1. Nerve pulses would not mediate information inside the brain. They would only build short connections between existing flux tube connections parallel to axons. Same happens in an old fashioned telephone network by relays: it would be energy consuming to keep the connections on all the time.

   The velocity of nerve pulse conduction is quite too slow to realize the iteration leading to a standardized sensory mental image. If the signal velocity is light velocity, duration of order 1 ms for nerve pulse also for 10 cm neural pathway about 10^6 forback travels between sensory cortex and retina.

   Communications would occur by dark photons signals with $h_{eff}/h = n$ and with maximal signal velocity allowing for an iteration leading to standardized perceptions as near as possible to the sensory input and representing only the essential features. Dark photons could transform in an energy conserving manner to biophotons with energies in visible and UV range (at least) and thus above thermal energy and therefore having effects not masked by thermal radiation. Brain is known to emit biophotons and they are also associated with axons [21, 20].

2. All information molecules (neural transmitters, hormones, messengers) would be connection builders so that the view of neuroscience would be badly wrong here. I have discussed this idea earlier but in a slightly different form: the proposal was that information molecules are attached to the end of a flux tube getting longer as the molecule travels to its target. This is possible but un-necessary since it is enough to build just the bridge between existing connections. **Remark:** The view of neuroscience might be very different if information technologies would have been known for a century ago. Same applies to homeopathy and water memory [18], which still remains curse words in mainstream science, although a lot about the mechanisms involved is known.

The standard view about learning as strengthening of synaptic connections would translate to a gradual build-up of permanent flux tube connections so that communications with dark photon signals would be possible all the time. This would lead to fusion of sender and receiver to a single quantum entangled system.

If the meridians of acupuncture network correspond to this kind of permanent network, they would not require nerve pulses, transmitters, nor information molecules.

3. Nerve pulse patterns would however generate Josephson radiation at EEG frequencies propagating from the brain to its MB from axonal membranes serving as Josephson junctions. EEG would code the nerve pulse patterns as frequency modulated Josephson radiation [17].

The view about sensory perception and function of nerve pulse transmission differs from the standard view. Nerve pulse transition would not be communication between parts of CNS but building of the communication line for dark photons making possible communications with maximal signal velocity [37][19].

1. This would allow generation of sensory mental images at sensory organs by an iteration involving virtual sensory input from brain to sensory organs. Pattern recognition would be realized as a build-up of an artwork representing standardized mental image as near as possible to the original sensory input.

2. Neurotransmitters and all information molecules would be bridges needed to construct connected communication lines. Learning as formation of permanent synaptic connections would be generation of permanent bridges of this kind.

3. Cell membrane and perhaps also other structures serve as generalized Josephson junctions [17]. The (generalized) Josephson radiation generated by nerve pulses would give rise to EEG (and perhaps also to its fractal counterparts) as communication of neural information from brain to MB via Josephson frequency modulation. The size scale of the layer of MB would be rather large, of the order $1/f_c$, of the order Earth size in alpha band ($f_c \simeq 10$ Hz).

This view allows to understand imagination as virtual sensory inputs *resp.* motor actions from MB via brain which do not reach actual sensory organs *resp.* muscles but virtual sensory organs inside brain for which a good candidates are basal ganglia - ganglions are also associated a with sensory receptors. Dreams (REM), hallucinations, and psychedelic experiences (motor activities during sleep) could be understood as virtual sensory input reaching the sensory organs (muscles).

Also memory recall could involve virtual (real in the case of sensory memories) sensory input from MB at which memory mental images are realized [52, 41].

3.2 Binaural beat as a support for TGD view about brain

The phenomenon known as binaural beat [8] provides support for the TGD view about the brain. Binaural beat occurs when sound waves with slightly different frequencies arrive in both ears. The beat can be understood as interference due to the time-varying phase difference of the waves. What is heard is the difference frequency, even when it is below 20 Hz - for instance 10 Hz-, and therefore not audible. The amplitude modulation with 10 Hz would be perceived, not the 10 Hz frequency. Strangely, the binaural beat occurs also when the signals arrive only to separated ears so that interference is not possible.

The TGD based explanation could be that the sound waves generate dark photon signals propagating along flux tubes and having classical em waves as correlates. The waves from different ears would interfere if the flux tubes meet at some point in the brain located at auditory areas perhaps. The first option is that this interference gives rise to the experience of the binaural beat and superposes with the sensory input assigned to ears (one cannot exclude the possibility that the sensory qualia are assigned to virtual sensory organs in the brain). Second option is that the virtual sensory input as feedback sent back to ears as dark photons superpose to the sensory input from ears.

3.3 The roles of nerve pulses and oscillations of neuronal membrane in the TGD picture

1. Nerve pulses - or more precisely, the transmitters emitted at synaptics contacts - connect flux tubes to longer pathways along which dark photons signals travel. Biophotons are dark photons transformed to ordinary so that there is empirical basis for this. Dark photons are an optimal tool for communications: light velocity and coherence.

   This allows the build of percepts as standardized sensory mental images by feedback. Nanosecond is the time scale for a single feedback loop so that there is a lot of time for this. This also explains dreams as virtual sensory input from the brain of MB to sensory organs in particular eyes (REM).

   Imagination can be understood as virtual sensory input which does not reach sensory organs or muscles but stops before it. Imagination is almost sensory experience with input from MB or brain. The notion of virtual sensory input is central for understanding speech comprehension and also inner speech.

2. Nerve pulses patterns modulate generalized Josephson frequencies for the membrane proteins (ion channels and pumps, etc...) and Josephson radiation to big bart of MB codes for the sensory input.

   Motor output is from MB in reverse time direction induced by BSFRs. A good guess is that it is via genes and induces gene expression by producing proteins but possible are also other forms of gene expression such as as dark photon signals to cell/neuronal membrane inducing nerve pulse patterns building connected wave guides for motor output as dark photons signals to propagate

3.4 Memories

To understand what memories and memory recall could be in ZEO one must specify what the geometrical correlate of subjective "Now" have?

1. The first proposal was that it corresponds to the active boundary of causal diamond (CD). It however turned out that the subjective "Now" could more naturally correspond to the $t = T$ slice of CD with maximal size located in the middle of the CD. Here t corresponds to a linear Minkowski time axis connecting the tips of the CD. If one accepts $M^8 - H$ duality [47], this picture can be made precise.

   The moments "Now" would correspond to "special moments in the life of self" [47, 51] identifiable as intersections of 6-spheres, which are brane-like entities (branes are encountered in M-theory) appearing as universal special solutions of algebraic equations determining the space-time surfaces in M_c^8. The values of T correspond to the roots of the real polynomial defining the space-time surface so that the values of "Now" are quantized.

2. During the sequence of state function reductions the active boundary of CD would shift towards the geometric future and the size of CD would increase (in statistical sense). The sub-CDs accompanying sensory and other mental images would shift to the direction of geometric future as CD increases and become potential memory mental images suffering BSFRs in a shorter time scale.

   The self would experience a memory mental image as a sub-self in memory recall to be discussed below. The time=constant snap-shots at the upper half of CD assignable to the memory mental images are ordered with respect to the Minkowski time t but the order is opposite to that for the subjective experiences. This was a great surprise to me. They would correspond to subselves to which memory recall builds a connection by entanglement quantally or by sending a signal, which is reflected back in BSFR for the memory mental images.

What about recall of episodic memories in ZEO?

1. Spontaneous memory recall could correspond to a death of a memory mental image with an opposite arrow of time and re-incarnation with the same arrow of time as self. This could be accompanied by emission of a past directed "negative energy" signal received by self associated with the moment "Now". The interpretation would be in terms of extraction of metabolic energy: memory recall indeed requires metabolic energy. Active memory recall could correspond to a receival of future directed "positive energy" signal coming from "Now" having interpretation as metabolic energy feed. Energy conservation would force the memory mental image to change the arrow of time.

2. The prediction would be that in active memory recall by a "positive energy" signal received by the memory sub-CDs, the order of recalled memories is opposite to that for the real experiences. There is evidence for this kind of change [10] (see also the popular article "*The human brain works backwards to retrieve memories*" at http://tinyurl.com/y7hbqmug).

3.5 Associations at quantum level

How associations could be formed at quantum level? Certainly memories and memory recall are involved and ZEO provides a universal model of memories.

1. In contrast to the naive expectations, in ZEO the memory mental images would be sub-selves and would comove with the active boundary of causal diamond (CD identified as an intersection of future and past directed light-cones) and shift to the direction of the geometric future after their creation at $t = T$ hyper-plane of CD at which upper and lower light-cones of CD are glued to together. This is the largest time slice of CD and assumed to define the geometric correlate for the subjective moment "Now".

   Memory mental image (associated with sub-CD) continues its Karma's cycle having as basic unit a birth in BSFR, a life consisting of a sequence of analogs of unitary time evolutions followed by SSFRs, death in BSFR and living a life with opposite arrow of time. Memory mental images can live in the brain of the geometry future being connected to the brain "Now" by long flux tubes.

2. Memory recall wakes up the memory mental images by sending a message using dark photons received by the memory mental image. The universal model of language suggests that the signal is biological system coded genes serving also as addresses.

3. Conditioning in its simplest form should associate two mental images. The classical example about conditioning is a dog, which learns to expect food after it hears the sound of a bell. The primary experience involves both the sound of the bell and getting the food. After the conditioning the mere sound of the bell stimulates activities like salivation. Positive or negative emotions facilitate conditioning. In ZEO framework the learning of the conditioned response would involve two mental images: imagined experience about obtaining the food and the sound of a bell.

   They should fuse to a composite mental image, perhaps by entanglement. These primary memory mental images and their almost copies produced later and involving only the bell and the imagined food would form a population of memory mental images in the geometric future shifting farther away. As the dog hears the sound of the bell, a message to the memory mental images in the geometric future is sent. It is realized as frequency modulated dark Josephson radiation from say basal ganglia of sensory organs.

4. A naive guess is that the modulated Josephson frequencies correspond to a period larger than the temporal distance of the memory mental image from "Now" and defining its age. Rather low frequencies are involved for long term memories and the values of h_{eff} must be scaled correspondingly. The longer the time span of the memory, the larger the value of h_{eff}. The emergence of language is therefore accompanied by the emergence of long term memory. The memory mental images about expectation of food +sound of bell have however a shorter time span. These signals wake-up the

memory mental images but they are however not conscious to self - and as they die they send a signal back to the brain inducing an imagined mental image involving also the promise of food.

5. In some cases the signal can reach the sensory organs and a sensory memory mental image is generated. This picture applies also to the acquisition of the language. The larger value of h_{eff} associated with language genes (the value of h_{eff} could vary for a given language gene) meaning larger layers of MBs and a possible fusion of MBs of he communicators, and therefore the ability to remember the associations of the words to sensory mental images for a long time. Hearing of the linguistic expression would also generate internal speech as a particular virtual motor action.

4 A TGD inspired vision about language

4.1 The role of MB

The proposal is that new layer of MB assignable to larger part of MB outside body was involved with the emergence of language. There are several arguments in favor of this proposal.

1. The model for how mutation of few genes like FOXP2 lead the evolution of human languages to be discussed relies on the idea that the value of h_{eff} assignable to dark variants of language genes increases. This means the emergence of new layer of MB having onion-like structure. What emerged was grammatics and syntax as hierarchical structures represented as many-sheeted space-time structures distinguishing humans from other animals could have emerged: these tructures can be assigned to MB and they have also interpretaton in terms of extension of rationals leading to n-sheeted structures. The new level of hierarchy would have emerged at the level of the MB including also dark gene first: flux tubes inside flux tubes inside labelled by values of h_{eff}.

2. The development of language led to a cultural evolution and could have been a quantum leap in the evolution of collective levels of consciousness: emergence of new levels in the hierarchy of extensions of rationals. Maybe the emergence of gene with large h_{eff} meant that it receives control commands from this collective level of consciousness possibly assgnable to communications, social group, or even larger structure. Recall that the size scale of MB assignable to EEG frequencies is of order Earth size. The basic structure of language are indeed very "social". Subjects, objects, verbs expressing what they do to each other, relations between these entities, attributes (adjectives) characterizing their states. Also the notions of plural and singular.

3. One can also ask how it is possible to distinguish between sensory input created by living beings and having meaning from that produced to dead matter. Also humans give emotional meaning to bird's song and vocal signals and gestures of animals but not to the sounds of dead Nature. For autists this ability might be very weakly developed. The natural answer to the queston is that all communications are also communications between magnetic bodies, quite concrete touching, makes it possible to distinguish natural sounds from speech and signals with represent communications. Communications require attention and the flux tube connections between communicators would be correlates for the attention.

    Mere mimicry does not require interpretation of the signal as communications. Some birds can mimic the sound of even a car. I remember my astonishment when finnish bird "talitiainen" mimicked the fate motif of Beethoven's symphony No. 5. My neighbours listened to classical music!

    There should be also a fundamental difference between the communications of ordinary sounds and speech to brain. The communications of speech could be via the large part of MB outside body whereas ordinary sensory data would be communicated via small part of MB to brain.

4. In language acquisition the role of parents, in particular mother, is crucial. One might of course argue that just mimicry and rewards are enough. But how the child knows that mother is trying to

teach her that the word "apple" corresponds to the object that the mother is holding at ther hand. Is the fusion and entanglement of MBs needed?

The acquisition of language by child might also involve the MBs of child and Mother at least fusing to a larger structure. This might help the child to understand that the purpose is to learn to reproduce the word associated with the object that word describes. It could also make possible to learn the grammatics and syntax by becoming a part of larger self already learned these notions.

5. Speech communications happen magically in a good company when people are friendly and benevolent. As a young man I was extremely shy in a company of people who were not my friends. When I had intention to say something, I tried to form sentences in my mind as internal speech before possibly getting courage to talk but found it extremely difficult and I remained usually silent. In a company of good friends I realized that it was not so difficult at all: someone talked through me using me as an instrument.

4.2 Genes and language

4.2.1 What is the role of FOXP2 and other control genes?

The question that led to the writing of this article was whether the mutation of the genome leading to FOXP2 gene and other similar genes responsible for control of the genome did lead to the evolution of human language. How? The above mechanism does not distinguish in no manner between linguistic and ordinary associations. What happened?

Evolution in TGD framework means the increase of number the increase of the complexity of extension of rationals and thus increase of its e dimension $h_{eff}/h_0 = n$ defining a universal measure of intelligence and also a measure for the temporal and spatial scale of quantum coherence. A possibly dramatic increase of h_{eff} for FOXP2 gene and other key genes is a natural hypothesis explaining why the complexity of the language evolved and led from signals to sentences requiring longer time scale of quantum coherence andalso emergence of complex hierarchical structures naturally assignable to the new extension as extension of the original one.

The larger the value of h_{eff}, the larger the scale of the layer of MB. This suggests that a new level of collective consciousness essential for communications emerged. This layer would be associated with the system formed by the systems communicating using language. This would explain the ability to distinguish between sounds produced by inanimate systems and sounds produced by living systems and having meaning.

The emergence of this new level would have meant emergence of many new things: of speech, of longer time scales of memory and planned action, of a new level of cognition, of imagination in longer time scales, and of cultural evolution.

Second mechanism relaed to the emergence of FOXP2 and other similar control genes could be energy resonance in the coupling of the analogs of DNA, RNA, tRNA, and amino acids. The coupling would be between the entire gene and its dark analog. Whether the energy resonance occurs for all cyclotron energies of codons separately or for their sum remains an open question. For both options small changes of the gene can spoil or produce an energy resonance. This sensitivity would make genes an ideal control tool but would also serve as a general mechanism also for genetic diseases. The increase of h_{eff} accompanied by a small mutation to guarantee energy resonance could be the mechanism explaining the importance of FOXP2 and similar genes.

4.2.2 What about the development of speech organs and brain areas related to speech?

The development of speech required development of speech organs and brain areas for understanding of language and language production. How important was their role or was the mutation of certain genes responsible for languagecontrol enough to initiate the evolution leading to the development of speech organs and needed brain areas?

Now I must again bring in a layer of MB - but with a considerably longer scale perhaps the scale possibly assignable to the entire species. MB indeed has layers or order of Earth size and much larger. The proposed emergence of a big layer of MB with a large value of h_{eff} could relate closely to Sheldrake's proposal about learning at the level of species. How this new layer could have affected the evolution of speech organs and new brain regions.

1. MB is the key player in TGD. The TGD Universe allows conscious entities and they tend to have ideas as we know. Did MB at some level of hierarchy get an idea about expression of information using temporal sound patterns coupling to dark photons with specific frequencies? That would be a representation of bio-harmony in a new much longer spatial and time scale: did this evolutionary step correspond to the emergence of a new even larger value of h_{eff} to the dark matter hierarchy. Maybe the realization of this new faculty would have been a fractally scaled up variant of earlier realizations making this easier. Did MB make a plan which was eventually realized after a lot of trials and errors?

2. What this plan could correspond to? Here p-adic physics enters into the game. p-Adic dynamics for p-adic variants of space-time sheets obeys the same field equations as real space-time sheets. It however allows breaking of a strict determinism of real number based field equations: this non-determinism would correspond to the freedom of imagination.

    p-Adic data could give rise to full space-time surface as dynamical patterns but they could correspond only to a piece of its real counterpart. Imagination would be non-realistic. Imagined motor actions and sensory inputs would correspond to this kind of partially fulfilled entions: signals would not reach sensory organs or muscles.

3. How this would apply to MB's plan to create sound producing organs? This plan could proceed by trial and error to become more realistic and gradually find a complete realization. The reduction of the planning to trial and error at dark gene level - would be an enormous simplification and could have meant mutations increasing the value of h_{eff} bringing in larger layers of MB related to the brain areas and speech organs.

4.3 Meaning from embodiment in the TGD framework

The notion of embodiment is central for the understanding of how speech gets its meaning. The simplest sentences represent sensory inputs or motor actions. But also very abstract expressions have metaphoric representations in terms of subject and objects and verbs representing actions. Embodiment means that language expressions are transformed to virtual sensory inputs and virtual motor actions creating imaginations of the real ones. This requires formation of associations as generation of sensory and motor mental images.

For instance, the sentence "A does something to B" creates virtual sensory and motor mental images in which A indeed does something to B. Mental images representing A and B and "does something" are generated and could correspond to interaction between two mental images. Basically remembering sensory percept in which A does something to B is enough to provide the meaning and the linguistic decomposition is a model. For instance, the heard speech generates internal speech helping in understanding. I wrote last year a TGD view about discovered coupling between the regions of the brain involved with hearing and production of speech.

The experience or imagined experience as virtual almost experience with input from MB rather than environment is associated with the expression of language. When the language has been learned, a mere language expression generates memory mental images about the experience associated with the expression. The mechanism is naturally pattern recognition and completion as a general mechanism of association and conditioning also in neuroscience and artificial pattern recognition.

4.3.1 Questions

In the TGD framework the questions are the following ones.

1. How memories are represented and how they give rise to conscious memory mental images? ZEO leads to a general proposal for how memory mental images are represented. First communication of sensory input to the part of MB containing a subself representing memory mental image, call it M. M receives the signals and experiences BSFR analogous to motor action involving a signal to the direction of geometric past to subself representing "Me Now". This signal is transformed to a nerve pulse pattern generating a virtual almost sensory mental image.

   The general proposal is that in biology at cellular level motor actions are generated as time reversed signals from MB to dark genome inducing neural activity by a signal to cell membrane. The signal from MB to genome would take place by dark photon representation of genetic code and induce BSFR. This mechanism would be quite general.

   Genes with N codons must be represented as a dark 3N-photon signal behaving like a single particle like entity. This is not possible in standard physics but adelic physics relying on number theory makes this possible. The notion of Galois confinement [49] makes possible dark photon 3N-plets representing genes as 3N-chords as units are analogous to baryons as color confined units formed from 3 quarks and thus behaving as dynamical units.

   The signal would generate a sequence cyclotron resonance peaks at the genome giving rise to a sequence of ticks at dark genome. They must in turn generate a signal to the cell membrane received as a sequence of ticks inducing the sequence of nerve pulses. This seems to require realization of genetic code at the level of the cell membrane level proposed [34]. The general principle would be the same as in LISP: only identical genes serving as addresses can be in communications by cyclotron resonance. Not only the notion of cyclotrotron radiation but also the notion of generalized Josephson radiation must be generalized: dark Josephson photons are replaced with dark 3N-photons.

2. Where the sensory signal to MB is generated? Its generation at neuronal or cell membranes as generalized Josephson radiation is not plausible since the time scales do not fit together. The modulation of Josephson radiation by nerve pulses patterns produces ripples rather than slow frequency modulation. A more plausible proposal is that the sensory signal to MB is generated at the basal ganglia of sensory organs as a generalized Josephson radiation with frequency modulation generated by the sensory input.

3. What is the basic quantum mechanism of association of the memory mental image B to a sensory input A? In the neuroscience framework it would happen in the associative regions of the brain by new pulse patterns and by learning based on changes in synaptic contacts. Now this would take place at analogous regions of MB to which sensory input is sent as a signal and induced cyclotron resonance for 3N-chords.

   A pattern recognition at the level of MB would be in question. This involves a completion of the sinput pattern - sensory mental image - to a pattern representing memory mental image associated with it. This requires a generalization of the existing vision about pattern recognition to quantum level. Also this step could involve resonance leading to a fusion of the associated mental images by entanglement. This fused pair of mental images would generate a dark 3N-photon signal propagating to the brain as a generalized cyclotron radiation.

4.3.2 Association to memory mental images gives meaning to linguistic expressions

1. Association as a manner to assign meaning to linguistic expressions by embodiment. Language expression is associated with an imagined sensory experience or motor activity. Also internal speech is imagined speech as imagined motor activity and generated by written text.

Association requires wake-up of memory mental image by the speech signal, which in turn generates a virtual sensory brain or lower level of layers of MB . In ZEO memory mental images are in the geometric future of "me Now" so that BSFR must take place: the memory self "dies" when it sends the message as a dark photon signal. The signal eventually arrives in the brain and generates a nerve pulse pattern needed by dark photon communications generating the virtual sensory to virtual sensory organs.

Memory mental images at MB are woken up in ordinary memory recall presumably taking place at the hippocampus. The frequencies involved are theta frequencies suggesting that the layers involved of MB have the size scale of Earth. In the case of speech the frequencies are in the range 150-300 Hz which suggests that layers corresponding to these frequencies are involved. Also longer time scales such as minute time scale are involved and much bigger layers of MB could be involved.

2. The signals could be sent to the MB from sensory organs.

   (a) Ganglions associated with sensory organs are analogous to brain nuclei and would be the primary receivers of the sensory input. Nerve pulses are generated by neurons above then. Ganglions must play an important role in the generation of sensory experience and motor activities. Ganglions in the brain are called basal ganglia. They could serve as receivers of virtual sensory input and motor output from the brain.

   The neuron structures above ganglions also generate nerve pulses and these give rise to communications to the brain along flux tubes associated with neural pathways by dark photons signals. These communications would represent ordinary sensory communications, in particular sounds as mere sounds without meaning. They would also give rise to language acquisition via association.

   (b) The view about communications to MB as Josephson radiation modulated by membrane voltage variations suggests that the frequency modulations of membrane potential at frequencies of speech are involved. The earlier proposal that nerve pulse patterns could induce this modulation. They however would correspond to ripples of long wavelength waves. Of course, also axonal membranes involve oscillations of the membrane potential inducing the modulation but this modulation of generalized Josephson energy involving also difference of cyclotron energies is much smaller than that caused by nerve pulses.

   The oscillations ganglion membrane potential induced by sound waves could be involved. Frequency modulated Josephson radiation modulated by sounds would propagate to some part of MB. One can consider even the possibility that dark genes such as FOXP2 generate dark 3N-photon radiation. These dark genes could be also realized at the level of cell membrane.

   What could be the radiation in the case of dark genes. Could it be generalized Josephson radiation assignable to an array of Josephson junctions defined by dark genes and their conjugates. Sound waves could induce frequency modulations of oscillations of the voltage between the dark genes just by putting them into motion. Does the distance matter.

   (c) The signals would be received by frequency resonance by some layer of MB responsible for memories representing word-sensory/motor associations. What this layer of MB is and where it is located? The flux tubes should allow 3-N dark photon sequences. Their realization outside the biological body does not look realistic. This suggests that the part of MB can be assigned with the brain of the geometric future. Magnetic loops would return back to the brain of the geometric future. The longer the time scale of the memory, the longer the loop. The realization of sensory or in part of MB analogous to associative cortex. What happens in the part of the MB of the future brain representing the memory about association. The analogy of pattern completion of incoming sound signal to sensory input should take place and generate a virtual sensory input to the geometric past as a response along flux loops arriving at the virtual basal ganglia defining virtual sensory organs. Two long loops would be involved. From sensory basal ganglia to the highest motor and sensory areas? And from these to virtual sensory and motor organs.

(d) The branching of axons suggests a branching of corresponding flux tubes. What could happen in this process? In branching the value of h_{eff} could be reduced for dark photons - for instance by frequency doubling. Frequency doubling would transform audible frequencies to patterns of nerve pulses with much higher frequencies. From long to short scales. h_{eff} hierarchy would be essential.

A possible interpretation as a cognitive quantum measurement is possible. Cognitive quantum measurement as a cascade of measurements in the group algebra of the Galois group of extension would give rise to a gradual reduction of effective Planck constants for the factors of the tensor product.

This cascade could correspond to the branching of axons leading to the reduction of biophoton energy in visible or UV to energy above thermal energy and assignable to cell membrane. What happens in branching of the flux tube? Is energy shared to that of n dark photons with the same frequency and smaller h_{eff}. Or does a localization to a single branch occur. h_{eff} would be reduced and f would increase. E would be conserved. Also both processes can occur. Division into n dark photons with $h_{eff} --> h_{eff}/n$ with f preserved plus a reduction $h_{eff}/n \to h_{eff}/nm$ and increase $f \to mf$ increasing by factor m.

(e) The communication via long flux loops to the small part of MB at the brain cannot correspond to this kind of process since the value of h_{eff} assignable to FOXP2 genes should be preserved. The communication could be to dark control genes such as dark FOXP2 generating signal to neuronal membrane - perhaps dark control gene also there - giving rise to nerve pulse pattern generating virtual almost sensory experience at the virtual sensory organs defined by basal ganglia.

This feedback should have been present already before the emergence of language but in shorter scales and leading to lower layers in the hierarchical structure of the brain ordered by evolution. They would correspond to a hierarchy of increasing values of h_{eff} realized at the level of genome.

These long feedback loops could end also at lower layers inside the brain and also the hierarchy of cortical layers could relate to this kind of feedback hierarchy. The virtual sensory input to the basal ganglia inside the brain would give rise to imagined sensory perceptions and motor actions.

(f) Interpretation as analog of Fourier transform is suggestive. The cyclotron resonance peaks would generate a sequence of ticks analogous to a Fourier transform of the incoming waves. Music-speech dichotomy suggests itself strongly. Speech could be analogous to a sequence of SFRs - ticks - and singing to superpositions of classical time evolutions connecting them. It has been said that the right brain sings and the left brain talks. Could the right brain sing in the sense that it would receive the signal as dark cyclotron radiation and could the left brain talk in the sense that this radiation would generate internal speech as virtual motor action.

A holistic representation in terms of frequencies would be transformed to "reductionistic" representation as time series. The correlation function for ticks would have the frequencies in its Fourier transform: stochastic resonance or its analog. Eventually this association to a sequence of ticks could generate a nerve pulse pattern creating a neural pathway making possible virtual sensory input in various sensory areas.

Given language expression corresponds to a huge number sensory percepts and one could argue that this requires a huge number of associations. In the computationalistic framework this would mean a huge amount of computer storage. The model for the generation of mental images predicts that the sensory mental images are standardized mental images generated by a feedback loop giving rise to a pattern recognition. Standard mental images allow also abstraction and conceptualization. One can even consider a quantum counterpart of the classical notion of concept. Concept as the set of its instances would be replaced by wave function in the set of instances giving a large number of different views about the concept.

4.4 Bio-harmony as a universal language

Bio-harmony [26, 27, 45] realizing genetic code for communications is an ideal candidate for a universal language: codon would represent 6 bits and the allowed 64 chords would represent mood at molecular level. There is quite a large number of fundamental moods. Both dark codons and 3-chords bound to units by Galois confinement [49] can be combined to dark genes by Galois confinement. This language would be minimal. The contents of the message would be minimal - the address of the receiver same as that of sender - so that LISP like language would be in question. The communications would be based on 3N-resonance. U-shaped flux tubes from receiver and sender forming bridges by reconnection would be the topological aspect of the communications.

The space-time surface associated with n:th order polynomial in M^8 defining the extension of rationals has n sheets corresponding to the roots of the polynomial [47, 41]. These many-sheeted structures would give rise to a geometric representation of hierarchical linguistic structures.

There is also an abstraction hierarchy defined by the functional composition of polynomials giving rise to representation of the Galois group of extension in terms of inclusion hierarchy of normal subgroups. Flux tubes within flux tubes within.... are possible. For extension of extension of ... with extensions having dimensions $n_1, n_2, ...$ one would have n_1-sheeted structure with sheets replaced with n_2 sheeted structures replaced with..... Substitution of x in $P_{n_1}(x)$ with $P_{n_2}(x)$ with x replaced with....would correspond to this replacement.

Cascades of quantum measurements for the states of the Galois group algebra to a product state in the tensor product of Galois group algebras of the hierarchy of normal subgroups would define cognitive measurements which could be crucial for understanding of language by analysis [L28].

4.4.1 Speech is only one form of communication of binary and emotional information

Concerning production and understanding of speech, one must see the situation more generally in TGD framework.

1. Speech is only one form to communicate information and emotions. Also gestures define a language being based on motor expression. An interesting test is how complex gestures developed before speech and whether FoxP2 has anything to do with sign language. Does sign language have grammatics and syntax characterizing formal languages? I remember a real life story about a person who expressed himself by precise dancing patterns: they weere like words or even sentences for him.

2. Music and singing is the second form of language and expresses emotions rather than bits. Here harmony is an essential notion. Some basic chords define the harmony expressing the mood. Bits/words do not matter, only the chords used.

    This leads in TGD to the model of bioharmony in terms of icosahedral and tetrahedral geometries and 3-chords made of light assigned to the triangular faces of icosahedron and tetrahedron. The surprise was that vertebrate genetic code emerged as a prediction: the numbers of DNA codons coding for a given amino-acid is predicted correctly. DNA codons correspond to triangular faces and the orbit of a given triangle under the symmetries of the bioharmony in question corresponds to DNA codons coding for the amino-acid assigned with the orbit.

    Codon corresponds to 6 bits: this is information in the usual computational sense. Bioharmony codes for mood: emotional information related to emotional intelligence as ability to get to the same mood allowing to receive this information. Bioharmony would be a fundamental representation of information realized already at molecular level and speech, hearing and other expressions of information would be based on it.

I have written a couple of articles about realization at molecular level inspired by the findings that RNA is central in conditioning and suggesting that RNA somehow represents emotions crucial for conditioning [35]. Dark DNA and bioharmony for which emotions are already realized would make it possible.

4.4.2 What does Universality mean?

There are two views about language: Universality (or computationalism involving only grammar and syntax) concentrates on the formal aspects whereas connectionism concentrates language as a conditioning. For the first option one speaks of language learning as learning of formal rules and this applies to written language and language of mathematics. For the latter option one speaks of language acquisition as an almost unconscious process of imitation. These two views would be fused together in TGD view.

1. There would be only one universal language at the fundamental level. For communications it would be defined by genetic code realized as 3-chords of dark photons forming in turn 3N-frequency composites serving also as units. This code has both the bitty aspect: codon corresponds to 6 bits and the emotional aspect defined by given bio-harmony characterizing that is by the 3-chords defining the bio-harmony and in this manner mood. Genome would define genotype of language and specific languages would be phenotypes.

   This code is used in communications between various levels of the hierarchy. At least in control commands arriving from MB to genome. The analog of Josephson radiation from cell membrane mediating sensory data to MB would consist of a sequence of notes but if cell membrane realizes genetic code, also Josephson radiation could consist of 3N-frequency dark photon composites representing genes. Note that the notion of tick makes sense also for 3N-chords. The message would be sent as Josephson radiation or cyclotron radiation and received as ticks corresponding to state function reductions.

   Of course, one cannot exclude the single note option - mere temporal pattern of ticks with varying time separations - for the messages to the genome could be the case of speech having constant pitch. For singing and speech mediating emotions the situation melody or sequence of 3-chords would be needed.

   Since the language would be realized at DNA level, even plants could communicate using it. Plants are known to communicate and there is evidence that plants can cognize and even count [2](https://cutt.ly/ffRYXH8). In TGD framework also hormonal communications thought to be chemical would take place by biophotons: the hormones connected by flux tube to molecule in say hypothalamus would build the waveguides to second molecule in body for dark photons to propagate.

   The basic new physics building bricks in this picture would be 3N-frequency cyclotron resonance transforming the oscillating signal from basal ganglia membranes to a sequence of ticks in turn inducing a sequence of nerve pulses generating the virtual sensory experience using stochastic resonance coding the frequencies of original signal to peaks in the frequency spectrum of the correlation function for the sequences of nerve pulses. Also dark 3N-photon Josephson radiation assignable to genes represented also at cell membrane level would emerge as a new concept.

2. The universal aspects of the language would be realized as a basic expression of dark genes realized in terms of 3N dark photon composites propagating along flux tubes. The content of the packet is the address to which it sent! This would be just like in computer language LISP. This would be the genotype of language, the universal language based on 3N-frequency-resonance between sender and receiver genes.

This would completely separate the meaning of language expressions from the basic communication mechanism. This is of course true also for kinds of communications. The sender and receiver provide the meaning for language expressions by sensory perceiving them. Understanding of how the meaning is generated is the key problem. This requires theory of consciousness and a new view about the conscious brain.

3. TGD view is based on dark 3N-photon resonance communications between genomes and possibly also the genomes associated with the cell membranes and microtubules realizing the genetic code. The sensory input together with the language expression would provide the primary sensory percept - just as in learning by example. When communicated to the brain and even MB a secondary virtual almost sensory percept and virtual almost motor action would be generated as imagined sensory inputs.

   This would be the fundamental association giving meaning to the language. Conditioning would occur and when the mere linguistic input is received, the virtual sensory precept and motor output are generated. Does this require anything new: for instance, does it require that the associations are remembered in some sense or are the associations realized as in neuroscience in terms of synaptic strengths? One would have memory as a learned behavior.

   First the sensory input generated by linguistic expression is communicated from the basal ganglia of sensory organ or virtual sensory organ to the sensory and motor cortices by using dark 3N-photon resonance. After this the virtual sensory input and almost imagined) perception is generated. How?: as dark 3N-photon signals propagating in opposite spatial direction to sensory organs. The fact that nerve pulse conduction is in a single direction only suggests that also time reversal occurs in BSFR.

4. This general picture applies to the formation of associations and conditioning quite generally. This would be also the mechanism of imagination, which also sharply distinguishes humans from animals. The special ability of the humans to imagine would have emerged at the same time as the complex language. This could be due to the mutations of certain language genes like FOXP2 acting as genes for which the 3N-photon resonance is realized and one must understand how this could be the case.

The proposed notion of universality is not in conflict with the fact there exist large number of languages. The development of different languages is actually easy to understand as reflecting the fact that there is underlying universal language which is minimal in the sense that the content of the message is the address of the receiver. Language acquisition is a conditioning process associating sensory inputs and motor outputs to language expressions at a more fundamental level and the words are just labels for them. This is like general coordinate invariance in general relativity. Points of space-time can have infinite manner of different labelings in terms of numbers (now words).

4.5 Geometrization and topologization of the grammar and syntax in terms of many-sheeted space-time

These aspects of speech make possible understanding of speech acquisition but what about intentional learning of speech involving learning of grammar and syntax, which have nothing to do with contents of speech? In computer languages and mathematics as language this aspect would dominate.

4.5.1 Fractal flux tubes networks and structures of language

The TGD proposal is that magnetic flux tube networks - possibly trees in case of speech and associated with nerve pulse patterns are in an essential role. Flux tubes are effectively 1-D and have orientation which corresponds to temporal direction of speech and spatial direction of written language. There are flux tubes inside flux tubes flux tubes giving rise to hierarchical structures corresponding to the parsing of language expressions. MB would as many-sheeted structure would geometrize/topologize grammar and syntax.

There are aso 2-D and even 3-D flux tube networks but not accompanied by neural networks. These would be essential for the geometric and holistic aspects of cognition. Visual cognition in particular. The meridian system of Eastern medicine would be associated with this. These flux tube networks would have been present before the emergence of the neural system and would be possessed even by plants.

Abstraction as thoughts about thoughts or functions of functions of ... Functional composition for polynomials in M^8 picture. h_{eff} hierarchy. Many-sheetedness. Infinite primes: hierarchy of quantum states. Multi-WCW. Abstraction of polynomial. Octonionic polynomial determined by a real polynomial with coefficients restricted to complex rational values. 2-variable case: X and Y. P(X,Y). Real pars of X and Y correspond to the same octonionic real axis otherwise independent. X restricted to the real axis. Octonionic roots.Discrete family. Wave function as an abstraction of a polynomial of degree n.Two many roots. Labeled by rationals. Sum does not converge.

TGD could reduce the structure of language to purely geometric structures. Sentences would correspond to many-sheeted space-time surfaces with their topology representing the parsing structure. Basic space-time sheets would represent words and by gluing them to larger space-time sheets one would obtain sentences. Non-associativity forcing use of brackets in mathematical expressions would be important. For instance. (AB)C would correspond to the structure formed from a pair A1C of space-time sheet at larger space-time sheet with AB topologically condensed at A1 replaced (A and B are flux tubs inside flux tube A1). A(BC) would correspond to AA1 with BC topologically condense to A1.

The hierarchy of extensions of rationals realized in terms of functional composition of polynomials defining space-time surfaces in M^8 as n-sheeted structures provides a number theoretical view about linguistic structures. The functional decomposition $P_1 \to P_1 \circ P_2(x)$ replaces each space-time sheet of the n_1-sheeted structure with n_2 sheeted structure associated with P_2.

4.5.2 How the structural elements of language can be understood?

One must understand what is behind the notions of subject, object, verb. How tempus, case, singular and plural , pronouns, adverbs, etc are expressed: at the level of genetic code or at the consciousness experience a contents of imagined sensory experience and motor activity associated with the experience.

Are tempus, case, singular and plural, attributes, pronouns, adverbs, etc coded already by the oscillation pattern of the basal ganglia membrane giving rise to imagined experience beside genuine sensory experience. This would be the most elegant option.

Same FoxP2 gene or its analogs could be involved. Consider tempus as an example. How the tempus would be coded to the oscillations of the ganglia membrane or to the position of these membranes in the brain - to what subself they represent. Who is talking and about what and when!

- "I see" would correspond to a real sensory perception.

- "I saw" corresponds to immediate personal memory: could this be a virtual almost percept produced by a memory and realized at different places as virtual sensory percept. Basal ganglia associated with a level higher than sensory organs responsible for imaginations and inner speech..

- "I will see" would correspond to sensory percept, precognitions in reversed arrow of time.

- "I have done" seems to refer to a remote past: different time scale and perhaps different value of h_{eff}.

- "I had done" is talk of another self above or parallel me in self hierarchy about me as sub-self as an outsider. Now the basal ganglia would be at some part of the brain containing mental images representing some outsiders, say community as sub-self.

One must also understand what makes a sentence a question or command. Again the associations between conscious experiences and language expressions should help. In written language formal tools to express whether the sentence represents a question, command or something else have emerged. Emoticons provide formal symbols for emotions.

5 Appendix: Living matter, biochemistry, and consciousness

The model for living matter relies heavily on the notions of MB carrying $h_{eff} > h$ phases behaving like dark matter and ZEO.

5.1 ZEO based quantum measurement theory extends to a theory of consciousness

ZEO based quantum measurement theory [52] leads to a quantum theory of consciousness (see **Fig. 1**) by lifting the observer from an outsider to part of physical system. In particular, the theory predicts that the arrow of time changes in "big" (ordinary) state function reductions (BSFRs) as opposed to "small" SFRs (SSFRs) as the counterparts of weak measurements (see **Fig. 2**).

This suggests that self-organization in all scales reduces to dissipation with reversed arrow of time. The energies of states increase with h_{eff} and h_{eff} tends to be reduced spontaneously. This means that energy feed is needed to preserved the distribution for h_{eff}: in biology this corresponds to metabolic energy feed. The energy feed necessary for self-organization would reduce to dissipation of self-organizing system in reversed time direction. Dark matter at MB of the system would serve as a master controlling the ordinary matter serving in the role of slave. Note that there would be master-slave hierarchy of MBs ordered by h_{eff}.

This would happen at magnetic and have dramatic implications. Time reversed dissipation looks like energy feed from the environment to system. Self-organization involves always energy feed and generation of structures rather than their disappearance in apparent conflict with second law. Self-organization would correspond to dissipation in reversed time direction implied by generalized second law. No specific mechanisms would be required and only metabolic energy storages- systems able to receive the energy dissipated in reversed time direction - are enough. Obviously this provides a totally new vision about energy technology.

5.2 p-Adic physics as a correlate of intention and cognition

One of the earlier ideas about the arrow of subjective time was that it corresponds to a phase transition front representing a transformation of intentions to actions and propagating towards the geometric future quantum jump by quantum jump. The assumption about this front is un-necessary in the recent view inspired by ZEO. Intentions should relate to active aspects of conscious experience. The question is what the quantum physical correlates of intentions are and what happens in the transformation of intention to action.

1. The old proposal that p-adic-to-real transition could correspond to a realization of intention as action. One can even consider the possibility that the sequence of state function reductions decomposes to pairs real-to-padic and p-adic-to-real transitions. This picture does not explain why and how intention gradually evolves increasingly stronger, and is finally realized. The identification of p-adic space-time sheets as correlates of cognition is however natural.

2. The newer proposal, which might be called adelic, is that real and p-adic space-time sheets form a larger sensory-cognitive structure: cognitive and sensory aspects would be simultaneously present. Real and p-adic space-time surfaces would form a single coherent whole which could be called adelic space-time. All p-adic manifolds could be present and define kind of chart maps about real preferred extremals so that they would not be independent entities as for the first option. The first objection is that the assignment of fermions separately to every Cartesian factor of the adelic space-time does not make sense. This objection is circumvented if fermions belong to the intersection of realities and p-adicities.

    This makes sense if string world sheets carrying the induced spinor fields- define seats of cognitive representations in the intersection of reality and p-adicities. Cognition would be still associated

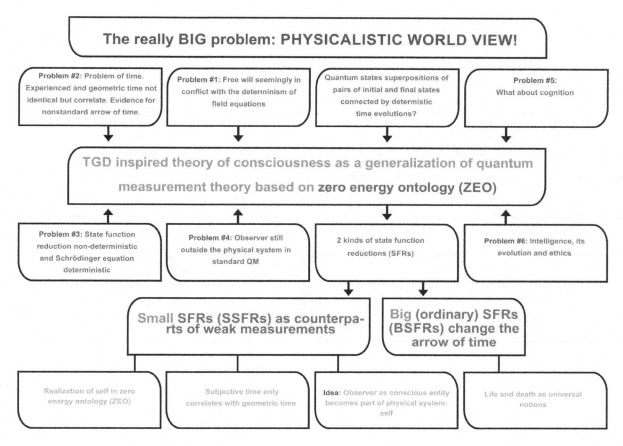

Figure 1: Consciousness theory from quantum measurement theory

with the p-adic space- time sheets and sensory experience with real ones. What can be sensed and cognized would be represented by the intersection.

Intention would be however something different for the adelic option. The intention to perform quantum jump at the opposite boundary would develop during the sequence of state function reductions at fixed boundary and eventually NMP would force the transformation of intention to action as first state function reduction at opposite boundary. NMP would guar- antee that the urge to do something develops so strong that eventually something is done.

Intention involves two aspects. The plan for achieving something which corresponds to cognition and the will to achieve something which corresponds to emotional state. These aspects could correspond to p-adic andreal aspects of intentionality.

The recent view relying strongly on $M^8 - H$ duality lead to ask whether the picture could be made more precise.This picture forces also to challenge the above picture.

1. The basic idea is that p-adic integration constants of the differential equation are pseudo-constants having a vanishing derivative but depending on finite number of pinary digits- rational numbers satisfy this condition. In M^8 picture a real polynomial with rational (or possibly algebraic) co- efficients determines the space-time surface. The roots of this polynomial as a function of radial

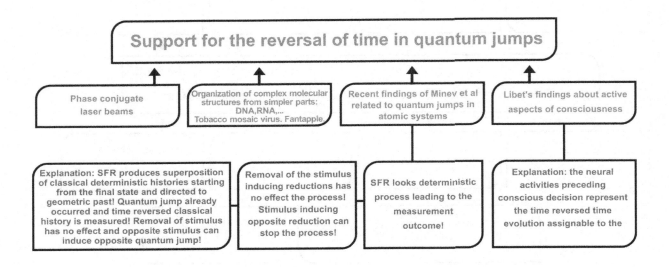

Figure 2: Time reversal occurs in BSFR

light-coordinate r at light-like boundary of CD determine this polynomial. When pseudo constant are allowed, the coefficients become pseudo constants, which are constants at the the interval $[0;T]$ divided to sub-intervals $I_1 = [0;t_1]$, $I_2 = [t_1;t_2]$, ..., $I_N = [t_{N-1};t_N]$ by the division $0 < t_1 < t_2 < ...t_N = T$.

2. Could the division to the intervals be unique by some argument? The roots of P are identified as moments for which SSFRs occur. Could tk correspond to a root of the polynomial P_k defined in the interva II_k. Could the "very special moments in the life of self" as roots of a polynomial correspond to introduction of new pseudo constants as a p-adic correlate for the state function reduction? Each interval has it own polynomial P_k and the allowed roots r_{k_i} become to the interval $[t_k;t_{k+1}]$ and their number is usually smaller than the degree n of the polynomial. Assume that each polynomial restricted to its own range defines a 4-surface inside the same CD. One would have m separate p-adic space-time surfaces. These surfaces would serve as correlates for intentions or dreams.

How could the real space-time surface as a realized intention relate to these surfaces?

1. Each of the 4-surfaces with genuinely constant coefficients of P_k has its own cognitive representation as points common to real and all p-adic variants. If the number of points t_k is finite one indeed has p-adic pseudo-constants for any prime p.

2. The realization of intention should be a quantum jump, state function reduction, or action of free will. Does this state function reduction have the selection of one of the polynomials P_k as a real polynomial defining the real space-time surface as a geometric correlate.

3. Could one generalize this to fermionic degrees of freedom. In [43, 44] I proposed that one could super-symmetrize TGD and quark spinors as imbedding space spinors by replacing imbedding space coordinates with super fields with components expressible as hermitian composites of second quantized quark and antiquark oscillator operators. Analogous generalization would be made for the second quantized quark field.

   In the M^8 picture the real polynomial would be replaced with a polynomial of super coordinate algebraically continued to super-octonionic coordinate. Solutions of the algebraic equations defining space-time surface would be now super-space-time surfaces which are unions of components assignable with the fermionic super coefficients of the super-polynomial.

   The rational coefficients of this polynomial could be replaced with pseudo- constants and the above picture seems to generalize. The spinor super-field would be a restriction of the M^8 spinor super-field to the p-adic branches of the p-adic space-time surface. Could the above picture about intentional act as a choice of the real branch generalize.

The next task is to understand intentional action at quantum level.

1. The most general vision is that intention corresponds to a superposition of p-adic spacetime surfaces with coefficients of polynomials which are genuine pseudo constants and by number theoretic universality same in all p-adic sectors. These superpositions would represent intentions and dreams. One could also speak of a dreamy CD containing a dreamy quantum Universe. Since cognitive representations are considered, everything would reduce to an extension of rationals, and the quantum dynamics by SSFRs and BSFRs would not formally differ from that for the real space-time surface and one could speak about transition amplitudes between dreams.

2. The realization of an intentional action would correspond to an SFR in which the pseudo constants become genuine constants. The simplest model is that one of the polynomials P_k is selected and be extended to a polynomial in the entire CD associated with P. The origin of CD is in a unique role in M^8 picture and $P(0) = 0$ makes possible hierarchies of extensions and conservation of number theoretical data as roots of P in the composition of polynomials realized for space-time surfaces.

   If $P_k(0)$ is required also for $k > 1$, any P_k can be selected. One can however challenge the idea that intentional action involves a selection. If $P_k(0) = 0$ for $k > 1$ is not assumed, P_1 associated with the interval $[0, t_1]$ must be chosen and CD corresponds to its size cale. One can talk about a partial realization of the intention in accordance with the intuitive expectations. For instance, imagined sensory percepts and motor actions could correspond to this kind of partial realizations.

3. If motor action corresponds to BSFR, intentional action can be realized only for BSFR. SSFR could not allow a realization of intention if the sequence of SSFRs corresponds to a functional composition of polynomials or even iteration of a single polynomial: I have considered these options for the sequence of SSFRs in [48].

4. This picture is in accordance with the conservation laws in ZEO and allows the creation of Universes as from nothing. CDs do not pop up from vacuum but dream-CDs transform to real ones.

One cannot avoid the question whether the notion of state function reduction could be reduced to a classical choice selecting one P_k: quantum jump as choice between dreams to be realized. This option would lead to purely classical probability theory and it would be however very difficult to understand what determines the transition probabilities.

5.3 The notion of magnetic body

Magnetic body (MB) would carrying dark matter would serve as the boss controlling ordinary matter at flux tubes.

1. MB has as building bricks magnetic flux quanta. Typically flux tubes and flux sheets. It consists of two kinds of flux quanta. Flux can be vanishing, which corresponds to Maxwellian case. The flux can be also non-vanishing and quantized and corresponds to monopole flux. In monopole case magnetic field requires no current to create it. This option is not possible in Maxwellian world. These flux tubes play a key role in TGD Universe in all scales.

2. Also Earth's magnetic field with nominal value $B_E = .5$ Gauss would have these two parts. Monopole part corresponds to the "endogenous" magnetic field $B_{end} = .2$ Gauss explaining strange effects of ELF em radiation to the physiology and behavior of vertebrates [6]. The presence of this part identifiable as monopole flux explains why Earth has magnetic field: this field should have decayed long time ago in Maxwellian world since it requires currents to generate it and they disappear. Magnetic fields of permanent magnets could have a monopole part consisting of flux quanta. Electromagnets would not have it.

3. MB would carry dark matter as $h_{eff} = n \times h_0$ phases and act as a "boss" controlling ordinary matter [46]. Communication to and control of biological body (ordinary matter) would be based on dark photons, which can transform to ordinary photons and vice versa. Molecular transitions would be one form of control.

4. Dark photons with large h_{eff} serve as as communication and control tools. Josephson frequencies would be involved with the communication of sensory data to MB and cyclotron frequencies with control by MB. Dark photons are assumed to transform to bio-photons [24, 23] with energies covering visible and UV associated with the transitions of bio-molecules. The control by MB which layers having size even larger than that of Earth means that remote mental interactions are routine in living matter. EEG would be a particular example of these communications: without MB it is difficult to understand why brain would use such large amounts of energy to send signals to outer space.

5. The experiments of Blackman and others led originally to the notion of h_{eff} hierarchy. The large effects of radiation at ELF frequencies could be understood iin terms of cyclotron transitions in $B_{end} = .2$ Gauss if the value of h in $E = hf$ is replaced with h_{eff}, which would be rather large and possibly assignable to gravitational flux tubes with $\hbar_{eff} = \hbar_{gr} = GMm/v_0$.

   MB would control BB by cyclotron radiation - possibly via genome accompanied by dark genome at flux tubes parallel to the DNA strands. Cyclotron Bose-Einstein condensates of bosonic ions, Cooper pairs of fermionic ions, and Cooper pairs of protons and electrons would appear in living matter and $h_{eff} = h_{gr}$ hypothesis predicts universal energy spectrum in the range of bio-photon energies.

   Cell membrane could act as generalized Josephson junction generating dark Jophson radiation with energies given by the sum for ordinary Josephson energy and of the difference of cyclotron energies for flux tubes at the two sides of the membrane. The variation of the membrane potential would induce variation of the Josephson frequency and code the sensory information at cell membrane to a dark photon signal sent to MB.

6. In ZEO field body and MB correspond to 4-D rather than 3-D field patterns. Quantum states are replaced by quantum counterparts of behaviors and biological functions. The basic mechanism used by MB would be generation of conscious holograms by using dark photon reference beams from MB and their reading. In ZEO also the time reversals of these processes are possible and make possible to understand memory as communications with geometric past. Sensory perception and memory

recall would be time reversals of each other and correspond to sequences of SSRs. Motor action would correspond to BSRs.

5.4 Life is not mere chemistry

The dogma about biology as mere bio-chemistry is given up in TGD framework.

1. Bio-catalysis remains a mystery in bio-chemical approach. MB carrying dark matter could provide the needed mechanisms.

    According to TGD view about catalysis, the U-shaped flux tubes associated with the MBs of reactants reconnect to a pair of flux tubes connecting the molecules [39]. This happens if there is cyclotron resonance for dark cyclotron radiation assignable to massless extremals (MEs) associated with U-shaped flux tubes. This requires that the flux tubes have same strength of magnetic field and therefore same thickness by flux quantization. The same value of h_{eff} guarantees resonance. The next step is the shortening of the flux tubes by a reduction of h_{eff} and liberating energy kicking the reactants over the potential wall making the process extremely slow otherwise.

2. Also valence bonds and hydrogen bonds could correspond to magnetic flux tubes characterized by $h_{eff} = h_{em} = n \times h_0$, where n is now rather small number ($h = 6h_0$). This leads to a model for valence bond energies of atom with n increasing as one moves to right along the row of the periodic table providing insights to the biological roles of various molecules in biology [32]. For instance, the molecules involving atoms towards right end of the periodic table would be natural carriers of metabolic energy whereas at the left end of row would be naturally involved with biocontrol via cyclotron frequencies.

3. The physics of water is full of anomalies [5]. TGD suggests an explanation [33] in terms of flux tubes assignable to hydrogen bonds [33, 38]. These flux tubes could correspond also to values of $h_{eff} > h$ so that these flux tube could be long and give rise to long range quantal correlations. Water could be seen as a manyphase system. The MBs assignable to water molecule clusters could mimick the cyclotron frequency spectrum of invader molecules and make possible water memory and primitive immune system based on reconnections of U-shaped flux tubes of water cluster and invader molecule [50]. In this framework water would represent a primitive life form.

    In Pollack effect [4] exclusion zones (EZs) are induced at the boundary between gel phase and water by energy feed such as IR radiation. The negative charge of EZs is explained as a formation of flux tubes carrying dark protons having interpretation as dark nuclei. A simple model for linear dark proton triplets predicts their states to be in 1-1 correspondence with DNA, RNA, tRNA, and amino-acids and the numbers of codons coding for given amino-acid are predicted to be same as for vertebrate genetic code [36, 45]. The model thus predicts deep connections between nuclear physics, condensed matter physics, chemistry, and biology usually thought to be rather disjoin disciplines.

    EZs are able remove impurities from interior in conflict with second law. TGD based explanation of the mystery is change of the arrow of time induced by TGD counterpart of ordinary state function reduction in ZEO) [52]: self-organization would be dissipation with reversed arrow of time at at the magnetic body (MB) of system acting as master and forcing time reversed evolution at the level of ordinary bio-matter serving as a slave.

    DNA has one negative charge per nucleotide, microtubules are negatively charged, also cell is negatively charged, ATP carries 3 units of negative charge. This together with ZEO suggests that Pollack effect plays a key role in bio-control and macroscopic SFRs play a key role in living matter.

Received February 23, 2021; Accepted October 1, 2021

References

[1] Smith C. *Learning From Water , A Possible Quantum Computing Medium*. CHAOS, 2001.

[2] Broberg A Anten NPR Ninkovic V Elhakeem A, Dimitrije Markovic D. Aboveground mechanical stimuli affect belowground plant-plant communication. *PLOS ONE*, 2018. Available at: https://doi.org/10.1371/journal.pone.0195646.

[3] Purves D et al. *What are basal ganglia?* 2001. August. Available at: https://www.neuroscientificallychallenged.com/blog/what-are-basal-ganglia.

[4] Zhao Q Pollack GH, Figueroa X. Molecules, water, and radiant energy: new clues for the origin of life. *Int J Mol Sci*, 10:1419–1429, 2009. Available at: http://tinyurl.com/ntkfhlc.

[5] Smirnov IV Vysotskii VI, Kornilova AA. *Applied Biophysics of Activated Water*. Word Scientific. Available at: http://tinyurl.com/p8mb97n., 2009.

[6] Blackman CF. *Effect of Electrical and Magnetic Fields on the Nervous System*, pages 331–355. Plenum, New York, 1994.

[7] Grigorenko EL. Speaking genes or genes for speaking? Deciphering the genetics of speech and language. *Journal of Child Psychology and Psychiatry*, 50(1G$_2$):116–125, 2009.

[8] Hink R et al. Binaural interaction of a beating frequency following response. *Audiology*, 19:36–43, 1980.

[9] Mozzi A et al. The evolutionary history of genes involved in spoken and written language: beyond FOXP2. *Nature. Scientific Reports*, 6(1):2–12, 2016. Available at: https://www.nature.com/articles/srep22157.

[10] Linde-Domingo and Wimber et al. Evidence that neural information flow is reversed between object perception and object reconstruction from memory. *Nature Communications*, 10(179), 2019. Available at:https://www.nature.com/articles/s41467-018-08080-2.

[11] Balter M. Speech Gene'Debut Timed to Modern Humans. *Science Now*, 6(22157):2–3, 2002.

[12] Chomsky N. *Aspects of the Theory of Syntax*. London: MIT Press, 1965.

[13] Kenneth P. *Language in Relation to a Unified Theory of the Structure of Human Behavior*. De Gruyter, 2015.

[14] Frank S. From molecule to metaphor: A neural theory of language (review article about the language theory of jerome a. feldman. *Computational Linguistics*, 33(2):259–261, 2007. Available at: https://cutt.ly/sfD7J3l. .

[15] Kempe V and Brooks PJ. *Modern theories of language (in Encyclopedia of Evolutionary Psychological Science)*. 2016. Available at: https://cutt.ly/jfvth1c. .

[16] Steven W and Slavoljub M. *Theory of Language*. A Bradford Book. London: MIT Press, 2000.

[17] Pitkänen M. Dark Matter Hierarchy and Hierarchy of EEGs. In *TGD and EEG*. Available at: http://tgdtheory.fi/pdfpool/eegdark.pdf, 2006.

[18] Pitkänen M. Homeopathy in Many-Sheeted Space-Time. In *Bio-Systems as Conscious Holograms*. Available at: http://tgdtheory.fi/pdfpool/homeoc.pdf, 2006.

[19] Pitkänen M. Quantum Model for Nerve Pulse. In *TGD and EEG*. Available at: http://tgdtheory.fi/pdfpool/pulse.pdf, 2006.

[20] Pitkänen M. Are dark photons behind biophotons? In *TGD based view about living matter and remote mental interactions*. Available at: http://tgdtheory.fi/pdfpool/biophotonslian.pdf, 2013.

[21] Pitkänen M. Comments on the recent experiments by the group of Michael Persinger. In *TGD based view about living matter and remote mental interactions*. Available at: http://tgdtheory.fi/pdfpool/persconsc.pdf, 2013.

[22] Pitkänen M. Nuclear String Hypothesis. In *Hyper-finite Factors and Dark Matter Hierarchy: Part II*. Available at: http://tgdtheory.fi/pdfpool/nuclstring.pdf, 2019.

[23] Pitkänen M. Are dark photons behind biophotons? *Journal of Non-Locality*, 2(1), 2013. See also http://tgdtheory.fi/pdfpool/biophotonslian.pdf.

[24] Pitkänen M. Comments on the Recent Experiments by the Group of Michael Persinger. *Journal of Non-Locality*, 2(1), 2013. See also http://tgdtheory.fi/public_html/articles/persconsc.pdf.

[25] Pitkänen M. Geometric theory of harmony. Available at: http://tgdtheory.fi/public_html/articles/harmonytheory.pdf., 2014.

[26] Pitkänen M. Music, Biology and Natural Geometry (Part I). *DNA Decipher Journal*, 4(2), 2014. See also http://tgtheory.fi/public_html/articles/harmonytheory.pdf.

[27] Pitkänen M. Music, Biology and Natural Geometry (Part II). *DNA Decipher Journal*, 4(2), 2014. See also http://tgtheory.fi/public_html/articles/harmonytheory.pdf.

[28] Pitkänen M. Cold Fusion Again . Available at: http://tgdtheory.fi/public_html/articles/cfagain.pdf., 2015.

[29] Pitkänen M. Strong support for TGD based model of cold fusion from the recent article of Holmlid and Kotzias. Available at: http://tgdtheory.fi/public_html/articles/holmilidnew.pdf., 2016.

[30] Pitkänen M. X boson as evidence for nuclear string model. Available at: http://tgdtheory.fi/public_html/articles/Xboson.pdf., 2016.

[31] Pitkänen M. DMT, pineal gland, and the new view about sensory perception. Available at: http://tgdtheory.fi/public_html/articles/dmtpineal.pdf., 2017.

[32] Pitkänen M. On the Mysteriously Disappearing Valence Electrons of Rare Earth Metals & Hierarchy of Planck Constants. *Pre-Space-Time Journal*, 8(13), 2017. See also http://tgtheory.fi/public_html/articles/rareearth.pdf.

[33] Pitkänen M. The Anomalies of Water as Evidence for the Existence of Dark Matter in TGD Framework. *Pre-Space-Time Journal*, 8(3), 2017. See also http://tgtheory.fi/public_html/articles/wateranomalies.pdf.

[34] Pitkänen M. About the Correspondence of Dark Nuclear Genetic Code and Ordinary Genetic Code. Available at: http://tgdtheory.fi/public_html/articles/codedarkcode.pdf., 2018.

[35] Pitkänen M. Could also RNA and protein methylation of RNA be involved with the expression of molecular emotions? Available at: http://tgdtheory.fi/public_html/articles/synapticmoods.pdf., 2018.

[36] Pitkänen M. On the Correspondence of Dark Nuclear Genetic Code & Ordinary Genetic Code. *DNA Decipher Journal*, 8(1), 2018. See also http://tgtheory.fi/public_html/articles/codedarkcode.pdf.

[37] Pitkänen M. DMT, Pineal Gland & the New View on Sensory Perception. *Journal of Consciousness Exploration and Research*, 9(3), 2018. See also http://tgtheory.fi/public_html/articles/dmtpineal.pdf.

[38] Pitkänen M. Two Poorly understood Phenomena: Maxwell's Lever Rule & Expansion of Freezing Water. *Pre-Space-Time Journal*, 9(5), 2018. See also http://tgtheory.fi/public_html/articles/leverule.pdf.

[39] Pitkänen M. How Molecules in Cells Find Each Other & Organize into Structures? *DNA Decipher Journal*, 8(1), 2018. See also http://tgtheory.fi/public_html/articles/moleculefind.pdf.

[40] Pitkänen M. An overall view about models of genetic code and bio-harmony. Available at: http://tgdtheory.fi/public_html/articles/gcharm.pdf., 2019.

[41] Pitkänen M. $M^8 - H$ duality and consciousness. Available at: http://tgdtheory.fi/public_html/articles/M8Hconsc.pdf., 2019.

[42] Pitkänen M. New results related to $M^8 - H$ duality. Available at: http://tgdtheory.fi/public_html/articles/M8Hduality.pdf., 2019.

[43] Pitkänen M. SUSY in TGD Universe (Part I). *Pre-Space-Time Journal*, 10(4), 2019. See also http://tgtheory.fi/public_html/articles/susyTGD.pdf.

[44] . Pitkänen M. SUSY in TGD Universe (Part II). *Pre-Space-Time Journal*, 10(7), 2019. See also http://tgtheory.fi/public_html/articles/susyTGD.pdf.

[45] Pitkänen M. An Overall View about Models of Genetic Code & Bio-harmony. *DNA Decipher Journal*, 9(2), 2019. See also http://tgtheory.fi/public_html/articles/gcharm.pdf.

[46] Pitkänen M. Self-organization by h_{eff} Changing Phase Transitions. *Pre-Space-Time Journal*, 10(7), 2019. See also http://tgtheory.fi/public_html/articles/heffselforg.pdf.

[47] Pitkänen M. New Aspects of $M^8 - H$ Duality. *Pre-Space-Time Journal*, 10(6), 2019. See also http://tgtheory.fi/public_html/articles/M8Hduality.pdf.

[48] Pitkänen M. Could quantum randomness have something to do with classical chaos? Available at: http://tgdtheory.fi/public_html/articles/chaostgd.pdf., 2020.

[49] Pitkänen M. New results about dark DNA inspired by the model for remote DNA replication. Available at: http://tgdtheory.fi/public_html/articles/darkdnanew.pdf., 2020.

[L28] Pitkänen M. The Dynamics of State Function Reductions as Quantum Measurement Cascades. *Pre-Space-Time Journal*, 11(2), 2020. See also http://tgtheory.fi/public_html/articles/SSFRGalois.pdf.

[50] Pitkänen M. Results about Dark DNA & Remote DNA Replication. *DNA Decipher Journal*, 10(1), 2020. See also http://tgtheory.fi/public_html/articles/darkdnanew.pdf.

[51] Pitkänen M. When does "big" state function reduction & reversed arrow of time take place? *Journal of Consciousness Exploration & Research*, 11(4), 2020. See also http://tgtheory.fi/public_html/articles/whendeath.pdf.

[52] Pitkänen M. Zero Energy Ontology & Consciousness. *Journal of Consciousness Exploration & Research*, 11(1), 2020. See also http://tgtheory.fi/public_html/articles/zeoquestions.pdf.

The TGD Based View about Dark Matter at the Level of Molecular Biology

M. Pitkänen[1] and R. Rastmanesh[2,3]

[1]Independent researcher. [1]
[2]Member of The Nutrition Society, London, UK.
[3]Member of The American Physical Society, USA.

Abstract

The notion of dark matter as phases of ordinary matter with effective Planck constant $h_{eff} = nh_0$ is the basic prediction of the number theoretic vision about Topological Geometrodynamics (TGD). This article is devoted to the possible role of magnetic body (MB) and dark matter in chemistry and biology. The first group of questions relates to the role of dark protons and electrons in ordinary and organic chemistry. Could the protons donated by acids be dark? What about the protons of hydrogen bonds? What about biologically important ions? What about oxidation and reduction: are the electrons involved dark: do valence electrons have $h_{eff} > h$? Second group of questions relates to the role of the MB carrying dark matter in biochemistry. Does the transition to biochemistry involve Pollack effect in which the fraction 1/4 of protons becomes dark and is transferred to magnetic flux tubes? Do dark protons form triplets forming analogs of basic biomolecules, which would be only secondary representations?

Dark protons could neutralize the phosphates of DNA and RNA. Could they also neutralize the phosphates at the ends of the lipids of the cell membrane: does the cell membrane realize genetic code? What about microtubules having GTPs associated with tubulins? ATP molecule has 3 units of charge: is it neutralized by dark proton triplet: could the energies of this triplet and dark valence electrons explain the high energy phosphate bond? Amino-acids should be accompanied by dark proton triplets: could dark electrons neutralize them? Dark photon triplets realize genetic code in terms of bioharmony. Could basic biomolecules and their dark analogs exchange dark photons in energy resonance? Could bio-photons result from dark extremely low frequency photons? Could energy resonance conditions select the basic biomolecules? These questions plus some long-lasting unsolved important questions will be addressed within the paper.

1 Introduction

The basic idea of the TGD based vision about living matter is that dark matter having effective Planck constant $h_{eff} = nh_0$ ($h = 6h_0$) located at the flux tubes of magnetic body controls ordinary matter: MB would be the boss and biological body the slave. This hypothesis can be justified by number theoretic vision about TGD, which unifies ordinary physics as physics of sensory experience described by real number based physics and the physics of cognition based on p-adic number fields: real and various p-adic number fields are fused to adele.

1.1 Physical motivations for the TGD notion of dark matter

The notion of dark matter as control of biomatter emerged before its number theoretic justification.

---

[1]Correspondence: Matti Pitkänen http://tgdtheory.com/. Address: Rinnekatu 2-4 A8, 03620, Karkkila, Finland. Email: matpitka6@gmail.com. Email: matpitka6@gmail.com.

1. The findings of Blackman et al [22] about the effects of ELF radiation (in EEG (electroencephalogram) frequency range) on vertebrate brain led to the hypothesis that besides protons also ions have dark variants having $h_{eff} = nh_0$ with $h_{eff} = h_{gr}$.

2. Also electrons could have these phases but now the value of h_{eff} would be much smaller and satisfy generalized Nottale hypothesis $h_{eff} = h_{em}$, where h_{em} is the electromagnetic analogue of h_{gr} assignable to flux tubes assigned with valence bonds. This leads to a model of valence bond (https://cutt.ly/5f5QrgF) predicting that the value of $h_{eff}/h_0 = n = h_{em}$ increases along the rows of the periodic table. This would explain why the molecules such as proteins containing atoms towards the right end of the rows serve as carriers of metabolic energy and why biologically important ions like C^{++} are towards the left end of the rows.

   The energy scale of dark variants of valence electrons is proportional to $1/heff^2$ so that the orbital radii are scaled up and the identification as a Rydberg atom is the only possibility in the standard physics picture: could dark valence electrons be in question? There is empirical evidence known for decades for the mysterious disappearance of valence electrons of some rare earth metals. The article "Lifshitz transition from valence fluctuations in YbAl3" by Chatterjee et al published in Nature Communications [4] discusses the phenomenon for Yb.

   The finding [5] about misbehaving Ruthenium atoms supports the view that covalent bonds involve dark valence electrons. Pairs of Ru atoms were expected to transform to Ru dimers in thermodynamical equilibrium but this did not happen. This suggests that valence electrons associated with the valence bond of Ru dimers are dark in TGD sense and the valence bonded Ru dimer has a higher energy than a pair of free Ru atoms. TGD based explanation [34] could be justified by a resonant coupling of the dark electron with an ordinary Rydberg state of the valence electron. In the lowest approximation dark valence electron has energies in the spectrum of ordinary valence electrons so that a resonant coupling with Rydberg states can be considered. The evidence found by Randell Mill [6] for atoms with an abnormally large scale of binding energy suggests the formula $h = 6h_0$ [33]. Adelic physics [31] predicts h_{eff} hierarchy and allows the understanding of the findings.

3. Nottale hypothesis [7] introduces the notion of gravitational Planck constant $\hbar_{gr} = GMm/v_0$ and is in the TGD framework identified as a particular value of h_{eff} assignable to gravitational flux tubes [37]. One trivial implication reflecting Equivalence Principle is that the cyclotron energy spectrum $E_c = n\hbar_{gr}eB/m = nGMeB/v_0$ does not depend on the mass m of the charged particle and is thus universal. The energies involved are proposed to be in the range of biophoton energies (at least) suitable for control of the transitions of bio-molecule.

The difference between non-organic and in-organic matter would be the presence of dark protons and electrons. The notions of acids and bases would reduce to the presence of dark protons: pH would characterize the fraction of dark protons. The notion of reduction and oxidation (REDOX reaction) would reduce to dark electrons associated with valence bonds (https://cutt.ly/5f5QrgF).

In biochemistry the density of dark protons would be much stronger and Pollack effect it in which the irradiation of water in presence of gel phases generates exclusion zones (EZs) as negatively charged regions by transferring every 4th proton to dark proton at flux tubes forming dark proton sequences as dark nuclei. Also dark ions become important in biochemistry, at least positively charged ions would have an important control role in TGD based view about biochemistry.

1.2 Realization of the vision about MB as controller of ordinary biomatter

$M^8 - H$ duality [39] concretizes the general vision. This duality states the representability of space-times as a 4-D surfaces in either complexified M^8 or $H = M^4 \times CP_2$. $n = h_{eff}/h_0$ has interpretation as dimension of extension of rationals and would the degree of a polynomial determining the space-time surface in M^8 as a root of polynomial of degree n. Roots would correspond to different sheets of n-sheeted space-time surface and Galois group of extension would permute the sheets with each other and act as a number theoretic symmetry group. Dark matter states at the flux tubes of B_{end} would be in representations of Galois group and Galois confinement [40] forcing n-particle states to behave as single unis like hadrons as color confined states.

The model of bio-harmony [27, 28, 38] based on the icosahedral and tetrahedral geometries in turn predicts that genetic codons correspond to dark photon triplets as 3-chords of lights. The representation of 12-note scale as a sequence of quints reduced by octave equivalence fixes the harmony for a given Hamiltonian cycle and realizes the symmetries of the harmony defined by some subgroup of the icosahedral group.

Combination of 3 icosahedral harmonies with 20 chords and having different symmetries with tetrahedral harmony with 4 chords gives bioharmony 20+20+20+4=64 chords assigned to DNA codons. Amino-acids are identified as orbits of 3-chords under the symmetries of a given harmony, and one obtains 20 amino acids. DNA codons coding for a given amino acid correspond to the chords at the corresponding orbit and the numbers of DNA codons coding for a given amino acid come out correctly.

Bio-harmony assigns the binary aspects of information to the 6 bits of codon and emotional aspects to the bio-harmony characterized by allowed chords fixed by a given Hamiltonian cycle at icosahedron and the unique tetrahedral cycle. The model of bio-harmony requires that the values of B_{end} correspond to those associated with Pythagorean scale and defined by quint cycle. These frequencies would correspond to energies that a molecule must have to serve as a candidate for a basic biomolecule.

In the model of genetic code [35] identifying codons as dark proton triplets, the numbers of dark proton triplets correspond to numbers of DNA, RNA, tRNA codons and amino acids and one obtains correctly the numbers of DNA and RNA codons assignable to given amino-acid in the vertebrate genetic code. Genes would correspond to sequences of dark proton triplets. Dark proton triplet would be analogous to baryon and Galois confinement[40] would force it to behave like a single quantum unit. Dark codons would in turn bind to Galois confined states of the Galois group of extension of the extension associated with the codons.

Galois confinement would be realized also for the dark photon triplets as a representation of genetic codons and also for the sequences of N dark-photon representing genes as dark 3N-photon states. Genes would serve as addresses in the communications based on dark $3N$-photon resonances. For communications between levels with the same value of h_{eff} there would be both energy and frequency resonance and for levels with different values of h_{eff} only energy resonance. It is an open question whether for dark-ordinary communications dark photon $3N$-plet transforms to single biophoton.

The basic hypothesis is that both DNA, RNA, tRNA, and amino acids are paired with their dark analogs, and that energy resonance mediates the interaction between the members of pairs. In this article the goal is to clarify the dark-ordinary pairing and the interaction between the members of the pairs. To achieve this, we first propose some questions below and then synthetize the answers to them.

1.3 Questions

In the sequel we will address the following questions about the roles of MB in the biochemistry of the basic biomolecules.

1. Do dark protons appear already in non-organic chemistry? Does acid/base tend to give/bind with a dark proton? The basic process is $OH \to O^- + H^+$. Water represents the basic example containing ions H_3O^+ and OH^-: the dark proton from H_2O would bind to the second H_2O acting in the role of base. pH characterizes the fraction of protons equal to 10^{-7} for $pH = 7$.

    Does the transition to biochemistry mean Pollack effect [21, 10] in which the fraction of dark protons becomes 1/4 corresponding to $pH = log_{10}(4)$. This would be the case for DNA, RNA, amino-acids, and tRNA also? Are the transitions between dark and ordinary states a key element of biochemistry. Could the gravitational flux tubes of MB take an active role in biochemistry?

2. Could the proton in hydrogen bond be dark? Could length of the hydrogen bond vary corresponding to different values of $h_{eff} = h_{gr}$. Could this explain the behavior of water below 100 C, in particular at physiological temperatures, challenging the standard thermo-dynamical model.

3. Do dark electrons play a role in chemistry as suggested (https://cutt.ly/5f5QrgF)? Does oxidation/reduction mean almost giving/receiving a dark valence electron in the valence bond? REDOX reactions are central also in biochemistry. The basic example is combustion in which O==O in presence of hydrocarbon such as sugar C_nH_{2n} gives rise to CO_2 and H_2O and $C_{n-1}H_{2n-2}$. O is reduced so that it almost receives valence electrons from C and H and C and H are in turn oxidized. The notion of electronegativity parametrizes the tendency to receive an electron. Is it possible to state that in inorganic and organic chemistry the electromagnetic part of MB is by far more important than the gravitational part of MB whereas in biochemistry also the gravitational part becomes important.

Also ions are proposed to appear as dark variants and one can wonder whether the valence electrons of positively charged biologically important dark ions like Ca^{++} are actually dark.

The following question can be asked about the role of MB in biochemistry of basic biomolecules.

1. Does the energy resonance for dark proton triplets and even for their sequences between biomolecules and their dark variants select the basic biomolecules like DNA, RNA, tRNA, and amino-acids having dark proton counterparts? Base pairs in the DNA double strand involve hydrogen bonds. Could these hydrogen bonds have dark variants?

2. Dark proton triplets would neutralize the negative charges assignable to the phosphates of DNA and RNA nucleotides and could be imaged as coming from POH$\rightarrow$ PO$^-$ +H$^+$ by a transformation of proton to dark proton by the analog of Pollack effect making DNA negatively charged.

What about the cell membrane, whose lipids have also phosphate ions at their ends? Could this give a higher level representation of the genetic code and genes at cell membrane level making possible dark 3N-photon communications between genome and cell membrane? Or do the dark protons serve at least as an energy storage? In fact, it has been proposed that cell membranes could involve a genetic code [20].

Microtubules are accompanied by negatively charged GTP molecules possibly associated with tubulins. 6-bit code defined also by DNA codons has been proposed by Hameroff et al as a memory code [23]. Could it be associated with genetic code represented using dark proton triplets?

3. The amino-acids in proteins should pair with dark variants of amino-acids by energy resonance. Amino-acid backbone does not however carry negative charge. Are the dark protons coming from NH_2 and COOH neutralized by electrons so that one would have dark hydrogens?

4. Also the ATP molecule has a negative charge of 3 units. Is it neutralized by a dark proton triplet serving as a temporary storage of metabolic energy? Could this energy at least partially explain the somewhat questionable notion of the high energy phosphate bond (also dark valence electrons would contribute)? Could ATP$\rightarrow$ ADP liberate metabolic energy by splitting one dark valence bond and transforming one dark proton to ordinary one? Do the dark protons assigned with the proteins serve as metabolic energy storage besides valence electrons, whose reduced Coulombic binding energies also give rise to higher than expected bond energies?

The next sections will be devoted to the possible answers to these questions.

2 Some number theoretical aspects of quantum biology

In this section the number theoretical aspects of TGD inspired quantum biology relevant to the recent article are considered. The role of the number number theory in TGD inspired view about cognition relying on adelic physics [31] is not discussed here.

Fig. 1 summarises the role of number theory in the TGD inspired vision concerning consciousness, cognition, and quantum biology and the role of dark matter in TGD inspired quantum biology **Fig. 2**

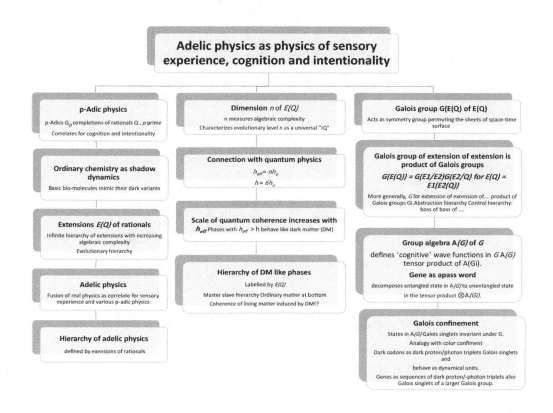

Figure 1: Adelic physics as physics of sensory experience, cognition and intentionality

2.1 Dark proton representation of genetic code

Fig. 3 summarizes the TGD based vision about genetic codes.

2.1.1 Codons as dark nucleons?

The model for codons of genetic code emerged from the attempts to understand water memory [24]. The outcome was a totally unexpected finding [35]: the states of dark nucleons formed from three quarks connected by color bonds can be naturally grouped to multiplets in one-one correspondence with 64 DNAs, 64 RNAs, 20 amino acids, and tRNA and there is natural mapping of DNA and RNA type states

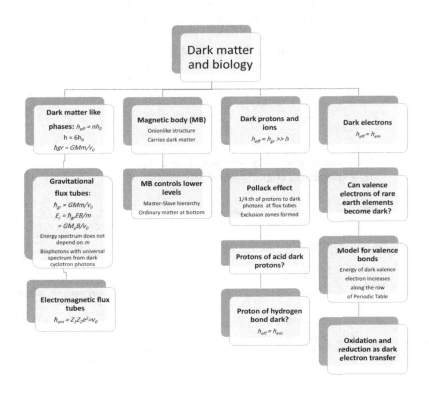

Figure 2: Dark matter in TGD inspired quantum biology

to amino acid type states such that the numbers of DNAs/RNAs mapped to given amino acid are same as for the vertebrate genetic code.

The basic idea is simple. The basic difference from the model of free nucleon is that the nucleons in question - maybe also nuclear nucleons - consist of 3 linearly ordered quarks - just as DNA codons consist of three nucleotides. One might therefore ask whether codons could correspond to dark nucleons obtained as open strings with 3 quarks connected by two color flux tubes or as closed triangles connected by 3 color flux tubes. Only the first option works without additional assumptions. The codons in turn would be connected by color flux tubes having quantum numbers of pion or η.

This representation of the genetic would be based on entanglement rather than letter sequences. Could dark nucleons constructed as a string of 3 quarks using color flux tubes realize 64 DNA codons? Could 20 amino acids be identified as equivalence classes of some equivalence relation between 64 fundamental codons in a natural manner? The codons would not be separable to letters but entangled states of 3 quarks anymore.

Genetic code would be defined by projecting DNA codons with the same total quark and color bond spin projections to the amino acid with the same (or opposite) spin projections. The attractive force between parallel vortices rotating in opposite directions serves as a metaphor for the idea. This hypothesis immediately allows the calculation of the degeneracies of various spin states. The code projects the states in $(4 \oplus 2 \oplus 2) \otimes (5 \oplus 3)$ to the states of 4×5 with the same or opposite spin projection. This would give

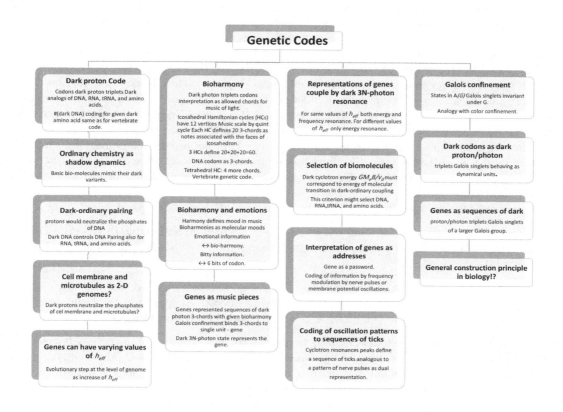

Figure 3: Genetic codes in TGD framework

the degeneracies $D(k)$ as products of numbers $D_B \in \{1,2,3,2\}$ and $D_b \in \{1,2,2,2,1\}$: $D = D_B \times D_b$. Only the observed degeneracies $D = 1,2,3,4,6$ are predicted. The numbers $N(k)$ of amino acids coded by D codons would be

$$[N(1), N(2), N(3), N(4), N(6)] = [2, 7, 2, 6, 3] .$$

The correct numbers for vertebrate nuclear code are $(N(1), N(2), N(3), N(4), N(6)) = (2, 9, 1, 5, 3)$. Some kind of symmetry breaking must take place and should relate to the emergence of stopping codons. If one codon in the second 3-plet becomes stopping codon, the 3-plet becomes doublet. If 2 codons in 4-plet become stopping codons it also becomes doublet and one obtains the correct result $(2, 9, 1, 5, 3)$!

2.1.2 Codons as dark proton triplets?

The model of codon as dark nucleon predicts analogs Δ resonances whose masses differ from those of nucleons.

The hint comes from the fact that DNA nucleotides have a negative charge, which is problematic from the point of view of DNA stability. This suggests that dark codons should have a charge of 3 units screening the charge of the ordinary DNA codon. Pollack effect [21] means formation of negatively charged exclusion zones as protons are transferred to dark protons at magnetic flux tubes. Could DNA be formed by Pollack effect? Could codons be represented as dark proton triplets?

The problem is that protons however have only 2 spin states: 4 states would be needed as in the case of quarks having also color. Where could the counterparts of spin and color come from?

One could consider adding a neural pion-like and/or ρ_0 meson-like bond connecting neighboring protons. Since ρ_0 has spin 1, this would give 1+3=4 states per bond. However, 2 states are enough and one must get rid of 2 states. The string-like structure of the proton triplet suggests that the rotation group reduces to $SO(2) \subset SO(3)$ so that ρ meson states split into singlets with helicities 0,1,-1. The doublet (-1,1) would serve as the analog of the isospin doublet (u,d) for baryons and enough to achieve a correct effective number $N = 4$ of states per single DNA codon. Helicity would replace isospin and the tensor product states could be constructed effectively as tensor products of 3 representations $2 \otimes 2$.

There is also an issue related to the fermionic statistics. Protons are fermions and the total wave function for them must be antisymmetric. For baryons color singlet property allows this. Can one require statistics in the ordinary sense also now? Or could the effective 1-dimensionality of the magnetic flux tube allow braid statistics?

The following variant gives good hopes about the ordinary statistics.

1. Adelic physics [31] brings in additional discrete degrees of freedom assignable to the group algebra of Galois group of extension of rationals inducing the extensions of p-adic number fields appearing in the adele [41].

2. Galois group acts on the space of space-time surfaces, and one can say that one has wave function at the orbit of the Galois group consisting of space-time sheets. At quantum level quantum states correspond to wave functions in the group algebra of Galois group of extension.

3. The role of color degrees of freedom in helping to achieve correct statistics in the case of baryon could be taken by Galois degrees of freedom. One can even consider the notion of Galois confinement as a generalization of color confinement [41] binding codons as dark proton triplets to dynamical units. Codons should be antisymmetric under exchange of dark protons in Galois degrees of freedom. Also genes as sequences of codons could be bound to dynamical units as Galois singlets. Could this allow ordinary statistics.

One can consider the replacement of u and d quarks with proton and neutron: color degrees of freedom would be missing but also now Galois confinement could come in rescue. Now however the charge screening of DNA by dark DNA would not be complete.

If this picture is correct, genetic code would be realized already at the level of dark nuclear physics or even at the level of ordinary nuclear physics if the nuclei of ordinary nuclear physics are nuclear strings. Chemical realization of genetic code would be induced from the fundamental realization in terms of dark nucleon sequences and vertebrate code would be the most perfect one. Chemistry would be a kind of shadow of the dynamics of positively charged dark nucleon strings accompanying the DNA strands and this could explain the stability of the DNA strand having 2 units of negative charge per nucleotide. Biochemistry might be controlled by the dark matter at flux tubes.

2.1.3 Cell membrane and microtubules as a higher level representation of genetic code?

Also the representation of genetic code at the level of cell membrane can be considered [35]. This kind of proposal have been made with different motivations by Okecukwu Nwamba [20]. The motivation for the current proposal is that the lipids have at their ends negatively charged phosphates just as DNA nucleotides have. The generalization of DNA as a 1-D lattice like structure to a 2-D cylindrical lattice containing nucleotide like units - letters - possibly assignable to lipids and realized as dark protons. Single lipid could be in the role of ribose+nucleotide unit and accompanied by a neutralizing and stabilizing dark proton. For axons one would have cylindrical lattice dark DNA lattice. The two lipid layers could correspond to two DNA strands: the analogs of the passive and active strand.

The finding is that membrane affects protein's behavior. This would be understandable in the proposed pictures 2-D analog of 1-D nucleotides sequences with codons replaced with counterparts of genes as basic units. That lipids are accompanied by phosphates with charge -1 gives the hint. Phosphate charge is neutralized by a dark proton as an analog of a nucleotide.

The notion of Galois confinement identifying genes as units consisting of N dark proton triplets representing genetic codons suggests that genes possibly assignable to the lipid layers of the cell membrane could communicate using dark $3N$-photon sequences with the proteins, genome, RNA and DNA. Dark variants of the control genes could initiate a nerve pulse pattern. An interesting possibility is that ganglions, nucleus like structures assignable to sensory organs and appearing as basal ganglia in brain [15] could communicate with genes.

Also microtubules have GTPs with charge -3 bound to tubulins. In dynamical instability known as treadmilling the transformation of GTP$\rightarrow GDP$ bound to β tubulin by hydrolysis induces the shortening of the microtubule at minus end whereas the addition of tubulins bound to GTP induces the growth at plus end. Also actin molecules bound to ATP show a similar behavior. Could they be accompanied by dark DNA codons? Are all codons allowed or does the absence of XTP, X= T,C,G mean that only codons of type GGG would be present?

For the dark codons for the cell membrane the p-adic length scale $L(151) \simeq 10^{-8}$ m would correspond to the lipid's transversal size scale and would be the distance between the dark protons. The scale of dark nuclear energy would be proportional to $1/L(151)$ and scaled down by factor $\sim 10^{-3}$ from that for DNA. The energy scale should be above the thermal energy at room temperature about .025 eV. If the energy scale is 2.5 eV (energy of visible photon) for DNA, the condition is satisfied. Note that 2.5 eV is in the bio-photon energy range. For p-adic large scales longer than $L(151)$ thermal instability becomes a problem.

It is interesting to compare the number of codons per unit length for ordinary genetic code (and its dark variant) and for various membranes and microtubules.

- For the ordinary genetic code there are 10 codons per 10 nm defining p-adic length scale $L(151)$. This gives a codon density $dn/dl = 10^3/\mu m$ in absence of coiling. The total number of codons in human DNA with a total length $L \sim 1$ meter is of order $N \sim 10^9$ codons. The packing fraction of DNA due to coiling is therefore huge: of order 10^6.

- If each lipid phosphate is accompanied by a dark proton and if lipid correspond to square at axonal cylinder with side of length $d = L(151)$ and the radius R of axon corresponds to the p-adic length scale $L(167) = 2.5\mu$ m (also of the same order as nucleus size), there are about $dn/dl = 2\pi(R/d)^2 \sim (2\pi/3) \times 10^4 \sim 1.3 \times 10^5/\mu m$. Axon should have length $L \sim 1$ cm to contain the entire genome.

  The same rough estimate applies to microtubules except that there would be one codon per GTP so that the estimate would be 3 times higher if GTP corresponds to length scale $L(151)$ of tubulin molecule. It has been proposed that genetic code is realized at the microtubular level.

- The nuclear membrane assumed to have a radius about $L(167) = 2.5\mu m$ could represent $N \sim (4/3)R^2/d^2 \sim .8 \times 10^5$ codons. This is a fraction 10^{-5} about the total number of codons. For a neuronal membrane with radius $R \sim 10^{-4}$ meters assignable to a large neuron the fraction would be roughly 10^{-1}. The fraction of dark codons associated with membranes could correspond to genes involved with the control and communication with genome and other cell membranes. Note that the non-coding intronic portion dominates in the genome of higher vertebrates. One can ask whether the chromosome structure is somehow visible in the membrane genome and microtubular genome.

2.2 Bio-harmony as a realization of genetic code

TGD leads to a notion of bio-harmony in terms of icosahedral and tetrahedral geometries and 3-chords made of light assigned to the triangular faces of icosahedron and tetrahedron [27, 28, 38]. The surprise was that vertebrate genetic code emerged as a prediction: the numbers of DNA codons coding for a given amino acid are predicted correctly. DNA codons correspond to triangular faces and the orbit of a given triangle under the symmetries of the bio-harmony in question corresponds to DNA codons coding for the amino acid assigned with the orbit.

Codon corresponds to 6 bits: this is information in the usual computational sense. bio-harmony codes for mood: emotional information related to emotional intelligence as ability to get to the same mood allowing to receive this information. bio-harmony would be a fundamental representation of information realized already at molecular level and speech, hearing and other expressions of information would be based on it. For emotional expression at RNA level possibly involved with conditioning at synaptic level see [36].

Does the generation of nerve pulse patterns by a gene mean at the cell membrane from dark DNA to dark protein map to dark protein (it could be also dark RNA or dark DNA even) associated with the cell membrane. What about communications with RNA and enzymes involved with transcription and translation. Do all basic biocatalytic processes involve them.

What about a generalization of Josephson currents? Dark ions certainly define them but could also dark proton triplets and their sequences associated with proteins give rise to oscillating Josephson currents through cell membrane and therefore to dark Josephson radiation with 3N dark photon units! Proteins themselves need not move much!

The universal language could be restricted to the genetic code which would be realized by dark proton triplets.The 64 codons are formed from 3 20-chord harmonies associated with icosahedron and the unique 4-chord harmony associated with tetrahedron. Bio-harmonies are associated with the so-called Hamiltonian cycles ,which go through every vertex of Platonic solid once. For icosahedron the number of vertices is 12, the number of notes in 12-note scale.

Also tetrahedron, cube, octahedron and dodecahedron are possible and one can consider the possibility that they also define harmonies in terms of Hamiltonian cycles. Dodecahedron would have 5-chords (pentagons as faces) as basic chords and there is only single harmony. Same mood always, very eastern and enlightened as also the fact that scale would have 20 notes.

Also octahedron gives 3-chords (triangular faces) whereas cube gives 4-chords (squares as faces). One can of course speculate with the idea that DNA could also represent this kind of harmonies: sometimes the 3N rule is indeed broken, for instance for introns.

Galois confinement [41] allows the possibility to interpret dark genes as sequences of N dark proton triplets as higher level structures behaving like a single quantal unit. This would be true also for the corresponding dark photon sequences consisting of 3N dark photons representing the gene in bio-harmony as an analog of a music piece consisting of 3-chords and played by transcribing it to mRNA.

The picture can be viewed even more generally. Any discrete structure, defining graph, in particular cognitive representation providing a unique finite discretization of space-time surface as points with the coordinates of the 8-D imbedding space coordinates in the extension of rationals, defines harmonies in terms of Hamiltonian cycles. Could also these harmonies make sense? The restrictions of the cognitive representations to 2-D partonic 2-surfaces would define something analogous to bio-harmony as Hamiltonian cycle of 2-D graph (Platonic surfaces solids can be regarded as 2-D graphs). The interpretation as representations of Galois groups and the notion of Galois confinement is possible although one loses the symmetries of the Platonic solids allowing to identify genetic code.

2.2.1 Brief details of the genetic code based on bio-harmony

TGD suggests several realizations of music harmonies in terms of Hamiltonian cycles representing the notes of music scale, most naturally 12-note scale represented as verticehttps of the graph used. The most plausible realization of the harmony is as icosahedral harmony [27, 28].

1. Icosahedron [1] has 12 vertices and Hamiltonian cycle as a representation of 12-note scale would go through all vertices such that two nearest vertices along the cycle would differ by quint (frequency

scaling by factor 3/2 modulo octave equivalence). Icosahedron allows a large number of inequivalent Hamiltonian cycles and thus harmonies characterized by the subgroup of the icosahedral group leaving the cycle invariant. This subgroup can be Z_6, Z_4, or Z_2 in order of reduced symmetry: Z_2 acts either as a reflection group or corresponds to a rotation by π.

2. The fusion of 3 icosahedral harmonies with symmetry groups Z_6, Z_4 and Z_2 gives 20+20+20=60 3-chords and 3+1 + 5 + 10 =19 orbits of these under symmetry group and almost vertebrate genetic code when 3-chords are identified as analogs of DNA codons and their orbits as amino acids. One obtains counterparts of 60 DNA codons and 3+1 + 5 + 10 =19 amino acids so that 4 DNA codons and 1 amino acid are missing.

3. The problem disappears if one adds tetrahedral harmony with 4 codons as faces of tetrahedron [2] and 1 amino acid as the orbit of the face of tetrahedron. One obtains 64 analogs of DNA codons and 20 analogs of amino acids: this harmony was coined as bio-harmony in [27, 28]. The predicted number of DNA codons coding for given amino acid is the number of triangles at the orbit of a given triangle and the numbers are those for genetic code.

4. How to realize the fusion of harmonies? Perhaps the simplest realization found hitherto is based on the union of a tetrahedron of 3 icosahedrons obtained by gluing tetrahedron to icosahedron along its face which is a triangle. The precise geometric interpretation of this realization has been however missing and some possibilities have been considered. The model could explain the two additional amino acids Pyl and Sec appearing in Nature [27, 28] as being related to different variant for the chemical counterparts of the bio-harmony.

There is also a slight breaking of symmetries: ile 4-plet breaks into ile triplet and met singlet and trp double breaks into stop and trp also leu 4-plet can break in leu triplet and ser singlet. This symmetry breaking should be understood.

2.3 Galois group of space-time surface as new discrete degrees of freedom

2.3.1 Galois confinemenent

The problem is to understand how dark photon triplets occur as asymptotic states - one would expect many-photon states with a single photon as a basic unit. The explanation would be completely analogous to that for the appearance of 3-quark states as asymptotic states in hadron physics - the analog of color confinement [40]. Dark photons would form Z_3 triplets under the Z_3 subgroup of the Galois group associated with corresponding space-time surface, and only Z_3 singlets realized as 3-photon states would be possible.

The invariance under $Gal(F)$ would correspond to a special case of Galois confinement, a notion introduced in [40] with physical motivations coming partially from the TGD based model of genetic code based on dark photon triplets.

2.3.2 Cognitive measurement cascades

Quantum states form Galois group algebra - wave functions in Galois group of extension E. E has in general decomposition of extension E_1 as extension of E_2 as extension of ... to a series . Galois group of E has decomposition to product of $Gal(E) = Gal(E/E_1)Gal(E_1)$ and same decomposition holds true for $Gal(E_1)$ so that one has hierarchy of normal subgroups corresponding extension of extension of...hierarchy defined by a composite polynomial $P(x) == P_1(P_2(x))$ with P_2 having similar representation. P defines in M^8 picture the space-time surface. This maps a tensor product composition for group algebra and the factors of group algebra entangle. SSFR corresponds to a cognitive quantum measurement cascade: SSFR in $Gal(E/E_1)$, SSFR in $Gal(E_1/E_2)$ etc.. The number theoretic measurement cascades for purely number theoretic Galois degrees of freedom are discussed in [41].

Could this cascade be analogous to the parsing of a linguistic or mathematical expression as cognitive measurements proceeding from higher to lower abstraction levels? Could the cascade correspond to a sentence S_1 about a sentence S_2 about ... such that one substitutes a concrete sentence for S_1 first, then to S_2, etc...? This is indeed suggested by the cascade of SSFRs since $h_{eff}/h_0 = n$ is the dimension of E_n.

Could cascade of flux tubes decaying to smaller flux tubes with smaller value of h_{eff} should correspond to this hierarchy. Certainly this is linguistics but the sentence as argument could correspond to several sub-sentences - different flux tubes. Could a neural pathway defined by the branching axon correspond to a concretization of this kind statement about statement (or multistatement, perhaps nerve pulse pattern generated by nerve pulse patterns arriving to a given neuron) about...

2.4 Energy and frequency resonance as basic elements of dark photon communications

Dark photon realization of genetic code leads to a view about fundamental linguistic communication based on resonance and we will write a separate paper connecting TGD with language soon. Two systems can be in communication when there is resonance. $E = h_{eff}f$ and energy conservation implies

$$h_{eff,1}f_1 = h_{eff,2}f_2 \ . \tag{2.1}$$

For $h_{eff,1} = h_{eff,2}$, energy conservation implies that both energies and frequencies are identical: $E_1 = E_2$ and $f_2 = f_2$. Both energy and frequency resonances in question.

In the general case one has $f_1/f_2 = h_{eff,2}/h_{eff,1}$ and frequency scaling takes place. The studies of water memory lead to the observation that this kind of phenomenon indeed occurs [11]. The communications of dark matter with ordinary matter and those between different values of h_{eff} involve only energy resonance. Frequency and wavelength scaling makes it possible for long scales to control short scales. Dark photons with EEG frequencies associated with the big part of MB transform to bio photons with a wavelength of say cell size scale and control dynamics in these short scales: for instance, induce molecular transitions. This is impossible in standard physics.

The resonance condition becomes even stronger if it is required there is a large number of biomolecules in resonance with dark matter realized as dark variants of biomolecules and dark ions. Cyclotron resonance energies are proportional to $\hbar_{eff}$ characterizing magnetic flux tubes and to the valued of the magnetic field strength dictated by the quantization of the monopole flux quantization by the thickness of the flux tube which can be do some degree varied by varying the thickness of the flux tube giving rise to frequency modulation.

The findings of Blackman et al [22] suggest that $B_{end} = 0.2$ Gauss defines an important value in the spectrum of B_{end} values. It could correspond to the field strength for the predicted monopole flux part of the Earth's magnetic field $B_E \simeq .5$ Gauss not allowed by Maxwell's theory. Besides B_{end} there would also be a non-monopole flux part allowed also in Maxwell's theory. Monopole flux part requires no currents as sources: this allows the understanding of the presence of magnetic fields in cosmological scales and also why B_E has not dissipated away long time ago [29].

There are however indications that the value B_{end} is quantized and is proportional to the inverse of a biologically important p-adic length scale and thus would be quantized in octaves. This could relate directly to the octave equivalence phenomenon in music experience. The model of bio-harmony [27, 28, 38] suggests a further quantization of the octave to Pythagorean 12-note scale of music. This would not be only essential for the music experience but communications of emotions and molecular level using the music of light.

2.4.1 Selection of basic biomolecules by energy resonance

The dark particles must have energy resonance with bio-molecules in order to induce their transitions. This seems to pose extremely strong conditions possibly selecting the bio-molecules able to form interacting networks with dark matter and with each other. One expects that only some amino acids and DNA type molecules survive.

Nottale's hypothesis provides a partial solution to these conditions. Nottale proposed the notion of gravitational Planck constant

$$\hbar_{gr} = \frac{GMm}{v_0} \qquad (2.2)$$

assignable in TGD to gravitational flux tubes connecting large mass M and small mass m and v_0 is velocity parameter. The gravitational flux tube presumably carries no monopole flux. The TGD based additional hypothesis that one has equals to

$$\hbar_{gr} = h_{eff} = nh_0 \ . \qquad (2.3)$$

This implies that the cyclotron energy spectrum

$$E_c = n\hbar_{gr}\frac{eB}{m} = n\frac{GM}{v_0}eB \qquad (2.4)$$

of the charged particle does not depend at all on its m. Therefore in a given magnetic field, say B_{end}, the cyclotron resonance spectrum is independent of the particle.

The energy resonance condition reduces to the condition that the charged ion or molecule has some cyclotron energy coming as a multiple of fundamental in its spectrum in the spectrum of its transition energies. Even this condition is very strong since the energy scale for cyclotron energy in B_{end} is in the bio-photon energy range containing energies in visible and UV. The fact that bio-photons have a quasi-continuous spectrum strongly suggests that B_{end} has a spectrum. The model of bio-harmony [38] suggests that the values of B_{end} correspond to Pythagorean scale constructible by quint cycle familiar for jazz musicians that is by taking $(3/2)^k$ scalings of the fundamental frequency and by projecting to the basic octave by octave equivalence.

The above simplified picture is formulated for single dark photon communications. The dark proton and dark photon realizations of the genetic code requires 3-resonance that is a simultaneous energy resonance for the 3 members of dark photon triplet. In dark-dark pairing also frequency resonance is possible. In dark-ordinary pairing frequency increases and couples long scales with short scales. Also resonant communications between genes with N codons involving $3N$ dark photon frequencies must be possible. This requires new physics provided by number theoretical vision.

2.4.2 What happens in the cyclotron resonance?

3 cyclotron energies for flux tubes characterize dark 3-proton triplet and Nottale's hypothesis predicts that they depend on the values of B_{end} for the flux tubes only. bio-harmony suggests that the spectrum of frequencies and thus B_{end} corresponds to Pythagorean 12-note scale for a given octave. The allowed chords of bioharmy would characterize the emotional state at the molecular level and correspond to the holistic emotional aspects of the communication beside the binary information.

The resonance would require that the dark cyclotron energy changes are equal to corresponding energies in molecular transitions. Galois confinement [40] makes possible also 3-N resonance. The resonance condition would select basic biomolecules and the ability of dark analogs of biomolecules to simultaneously resonate with several biomolecules would give additional conditions. In particular this would select DNAs and amino acids.

An open question is whether the coupling to ordinary biomolecules involves a transformation of a dark photon triplet or an N-plet to a single ordinary photon. For instance, does the sum of the 3 cyclotron excitation energies appear in the coupling of dark 3-proton state to amino acid in protein? This would have an analog as 4-wave coupling [3] in laser physics allowing in biology the transformation of dark photon triplet to single biophoton/or 3 bio-photons or *vice versa*. 6-wave coupling of laser physics would be analogous to the coupling of ordinary 3-photon state to dark 3-photon and back to ordinary 3-photon state.

The resonance itself would mean a process in which dark 3-proton cyclotron excitation returns to the ground state and generates dark 3-photon transforming transforming to ordinary photon (or 3-photon) and absorbed by the ordinary codon or amino acid excitation to hither energy state. This state would in turn emit an ordinary photon transforming to dark 3-photon absorbed by dark codon. This mechanism generalizes to 3N-proton states representing genes or dark proteins.

3 Some applications

3.1 How to understand the pairing between basic biomolecules and their dark variants?

There are interesting questions concerning the analogs of transcription and translation. Could dark DNA send signals also to dark RNA and amino acids and dark RNA to dark amino acids and dark tRNA? Could 3-photon resonance make it possible for biomolecules to find each other in the molecular crowd as proposed. This would be possible when the moods (bio-harmonies are the same - only an unhappy person can really understand an unhappy person!). For genes the 3-flux tube would be replaced with 3N-flux tube made possible by Galois confinement [40] .

3.1.1 Where do the dark proton sequences associated with proteins come from?

In the formation of protein 3 dark protons drop to a larger space-time sheet. The charges of amino acid residues vary in sign, vanish, or they are neutral, polar, or non-polar. Therefore the dark proton triplets must somehow be associated with the protein backbone as they do in the case of DNA and RNA. This implies that it is ionization of acidic groups OH (as in case of phosphates in DNA) or NH_2. The pairing with the residues would come from 3-photon cyclotron resonance.

Where do the dark protons come from? The backbone of protein is in the same role as sugar phosphate backbone in DNA and RNA. Amino acid residues are in the same roles as DNA nucleotides.

1. Amino acids are acids: NH_2 and COOH groups make them acidic. They tend to release protons and become negatively charged. They could give dark protons. In the formation of protein $NH_2 \rightarrow$ NH: one proton and electron lost. Does the proton come dark?

    Where does the electron go? Is it also dark and bound with a dark proton to form a dark atom? This kind of option in the case of the TGD based model for cold fusion [30, 32] involving dark dark proton sequences in a smaller scale.

2. C-OH loses H as C-O-N is formed. Both electron and proton are lost. Also this proton could become dark and bind with the dark electron to form dark hydrogen atom.

3. Where does the third dark proton come from? Is also NH in C-NH of the peptide acidic? Can it lose a proton, which becomes dark? Just as in the case of DNA codon, electrons would neutralize the dark proton. One would have instead of a dark proton sequence a dark H sequence. The additional charge of amino acid can be positive or negative and its possible polarity relates to the residues and to chemistry. The backbone would serve as the interface between dark matter and chemistry. The resonant interaction between the dark amino acid and residue would give the pairing between amino acid and its dark counterpart.

3.1.2 Denaturation of proteins and DNA

One can wonder how the denaturation of proteins and DNA could relate to dark protons.

1. Do the dark hydrogens become ordinary in the case of protein? h_{eff} would be reduced and the protein would decay. The energy liberated from dark protons and be used to store metabolic energy in the catabolism of proteins.

2. In the denaturation of DNA double strand hydrogen bonds between strands are lost. This also happens in DNA strand opening during transcription and translation. This cannot relate to a loss of dark proton sequences, which would lead to depolymerization.

    Why does the loss of hydrogen bonds lead to the denaturation? Is there binding between dark codon sequences inducing the formation of hydrogen bonding? Is Galois singletness for Z_3 replaced with Z_6 singletness so that a bound state of 2 dark proton triplets corresponding to codon and conjugate would be formed: this would be codon pairing at the level of the dark genome. This is considered in [40].

3.1.3 Hydrogen bonds and energy resonance

If also hydrogen bonds involve dark proton, there should be an energy resonance in which the dark proton returns from an excited cyclotron state and gives energy to the molecule to which it is bound and excites it. This would then decay to ground state and give the energy back to the dark proton. This would be kind of quantum tennis.

Hydrogen bonds would be also present between the paired bases: depending on the base pair their number would be 2 or 3. These dark protons would not correspond to those associated with dark DNA strands. An interesting question is how important the pairing of dark DNA strands and analog of hydrogen bonds of base pairs is and whether it relates to the energetics assigned with hydrogen bonds.

For instance, one can ask why A-T pairing by hydrogen bonds rather than A-C pairing is good.

1. Suppose that the dark codons DA and DT have the same 3-frequency giving rise to frequency resonance between them so that they can pair. DA and DC do not have the same 3-frequency and cannot pair. Pairing is therefore unique at the dark level.

2. The energy resonance condition assigns to a dark codon a unique codon so that one obtains only A-T pairing induced by dark pairing.

3.2 Does high energy phosphate bond involve 3 dark protons?

High energy phosphate bond plays a key role in the modelling of ATP hydrolysis [8] in the framework of standard chemistry. The official view is that everything is well-understood but for instance Ling has criticised both the notion of the high energy phosphate bond and the reduction to the molecular level [18, 19, 13, 16, 17] and also emphasised the importanec of a network like structures assignable to the cellular water: in TGD these networks would relate to MB. The work of Ling is discussed from TGD point of view in [25, 26].

From the TGD point of view the notion of high energy phosphate bond would be a mistake at the level of fundamental physics: dark matter and MB would be neglected. Thermo-dynamical chemistry can cope with this phenomenologically by introducing the notion of chemical potentials effectively describing the presence of dark matter. What is lost is quantum coherence in longer than atomic scales needed to really understand life.

The energy carried by 3 dark protons should replace the notion of high energy phosphate bond. Pollack effect indeed requires energy feed and this energy would go to dark protons taken from water.

Remark: Pollack effect would be an extreme example of adicity. Every fourth proton would become dark proton at flux tubes. pH would be $log_{10}(4)$! Also ordinary acidity could mean presence of dark protons but their number would be extremely small: one has fraction 10^{-7} for $pH = 7$.

This view would hold also more generally. The dark protons associated with proteins would also serve as a metabolic energy storage. In the denaturation this energy would be liberated. This happens in composts in which the organic material decays and causes heating of the compost. Of course, also the valence bonds which are dark carry energy as energy of dark electrons: by $h_{eff} > h$ the Coulombic bindin energy would be reduced and the energy of the valence bond would increase.

ATP→ADP and also ADP→ AMP [12] are possible. Dark electrons associated with the valence bonds could contribute to bond energy since large $\hbar$ reduces the negative Coulomb interaction energy assignable to the bond.

Also the dark protons associated with the phosphates could contribute the energy assigned usually with high energy phosphate bond. Pollack's finding [21] about the formation of exclusion zones (EZs) in presence of irradiation, most effectively IR radiation, led to the TGD based model. A considerable fraction of protons (fraction of 1/4) would be transferred to dark protons at the dark flux tubes. This requires (metabolic) energy and IR radiation would provide it and the energy is stored as energy of dark protons. Hence the pure chemistry based view about high energy phosphate bond would be wrong.

ATP has three phosphates and negative charge of -3 units. It would be screened by charges of 3 dark protons at the flux tube associated with ATP defining possibly a dark DNA codon (adenine triplet?). Dark RNA is not allowed since RNA does not allow A but U instead of it. In ATP → ADP the energy is given as a photon to the enzyme catalyzing the reaction allowing to overcome activation energy barrier. In microtubules one has GTPs binding stably to α tubulins but not β tubulins.

Microtubules (MTs) define an interesting candidate for the realization of genetic code. One can also try to understand MTs and their dynamics in terms of Galois confinement.

1. The model of 6-bit memory code [23] discussed by Hameroff et al relies on the hexagonal lattice formed by tubulin dimers consting of a pair of α and β tubulins, the 6-foot structure of CaMKII kinase domains, and the fact that the hexagon and CaMKII fit nicely together. The dynamical tubulins must be β tubulins for which the phosphorylation is not stable. The phosphorylation state of a given foot of the CamKII kinase domain represents a single bit so that CaMKII stores 6 bits. Its attachment at the hexagon of 6 tubulin dimers containing one tubulin dimer at is center could transfer the GTPs and thus 6 bits of information to the center tubulin. The proposed interpretation is as a transfer of information from neuronal to microtubular level involved with the synaptic learning.

2. The TGD inspired question is whether the CaMKII kinase domains are accompanied by dark proton triplets transferred to the tubulin dimer at the center of the hexagon so that microtubules would provide a 2-D representation of genetic code. If CamKII affects only the dark codon at the center of the hexagon, the center hexagon can behave as indepnt 6-bit units making possible 2-D lattice representation of the genetic code. This framework does not allow charge neutrality, and microtubules are indeed negatively charged having positively charged and negatively charged ends. Second option would be that the stable GTPs associated with α tubulins define an analog of genome with single codon per GTP.

3. GTPs at the minus end of MT stabilize it, and GTP → GDP transition liberating energy occurring for β tubulins causes the thread mill instability illustrated by a video of the Wikipedia article [9] about MTs. The 13 linear strands of tubulin dimers separate and bend radially outward. Are the 13 tubulin strands Galois confined states of tubulin dimers? Do the 13 strands form a Galois confined state as well? Does the liberated energy overcome the activation energy barrier against the decay to 13 separate tubulin strands?

   The video of the Wikipedia article [9] illustrates the formation of the structure. Could the decay correspond to a cascade of cognitive measurements leading from a state in Galois group algebra to an entangled product state in the tensor product of states assignable to the group algebras of normal Galois subgroups associated with an extension of extensions of ... of rationals [41].

3.2.1 The energetics at the dark proton flux tube

The energetics of the flux tube containing 3 dark protons must be considered.

1. Consider first the Coulombic interaction energy between dark protons. The interaction energy includes Coulombic interaction energy of nearest neighbor dark protons with distance R and those with distance of 2R.

   (a) If the flux tube is open, then we have

   $$E_c = \frac{2e^2}{R} + \frac{e^2}{2R} = \frac{5}{2}\frac{e^2}{R} \equiv \frac{5}{2}E_0. \tag{3.1}$$

   (b) If the flux tube is closed one has

   $$E_c = 3E_0 . \tag{3.2}$$

2. There is also strong interaction energy (one has a dark nucleus). Strong interaction is short ranged.

   (a) If the flux tube is open one has strong interaction energy $2E_s$ and total energy is

   $$E_{open} = \frac{5}{2}E_0 + 2E_s . \tag{3.3}$$

   (b) If the flux tube is closed one has

   $$E_s = 3E_0 + 3E_s . \tag{3.4}$$

3. There is also the total negative Coulomb interaction energy of dark protons with the total charge of phosphates.

   $$E(c, N) = K(N)E_P , \tag{3.5}$$

   where E_P is interaction energy between dark proton and phosphate. $N = 3, 2, 1$ for ATP, ADP, AMP If the dark protons interact as independent entities with 3 different phosphates one has $K = N$. If both ATP and protons act as single charged entities this energy one has $K = N^2$

4. The total energies for ATP, ADP, AMP are given by

   $$\begin{aligned} E_{open,3} &= \tfrac{5}{2}E_0 + 2E_s + K(3)E_P , & E_{closed,3}(ATP) &= 3E_0 + 3E_s + K(3)E_P . \\ E_{open,2} &= E_0 + E_s + K(2)E_P , & E_{closed,3}(ATP) &= E_0 + E_s + K(2)E_P . \\ E_{open,2} &= K(1)E_P , & E_{closed,3}(ATP) &= K(1)E_P . \end{aligned} \tag{3.6}$$

   where $K = N$ or $K = N^2$. Note that for $N = 1$ there is no difference between open and closed cases.

5. What happens in ATP→ADP and ADP→AMP? One organic phosphate (P) transforms to inorganic phosphate ion P_i without dark proton and one dark proton is lost. There are two left. Energy is liberated. There are also other contributions but let us forget them for a moment. The energy liberated is $E \simeq .5$ eV, metabolic energy quantum, energy of an IR photon.

The liberated energy is in various cases

$$\begin{aligned} \Delta E_{open}(ATP \to ADP) &= \tfrac{3}{2}E_0 + E_s + [K(3) - K(2)]E_P \ , \\ \Delta E_{closed}(ATP \to ADP) &= 2E_0 + 2E_s + [K(3) - K(2)]E_P \ , \\ \Delta E(ADP \to AMP) &= E_0 + E_s + [K(2) - K(1)]E_P \ . \end{aligned} \quad (3.7)$$

3.2.2 Empirical input

The reconstruction of ATP requires 1 dark proton and free energy about $\Delta G = -.5$ eV is needed. Actually 3 or 4 protons arriving through the cell membrane and getting kinetic energy in the membrane potential are used. Where does the surplus energy go? Or is there any surplus energy at all?

1. Mitochondrial membrane potential for proton which is determined by Coulomb potential and chemical potential due to the proton concentration difference at two sides of the membrane is about .15 eV [14]. Multiplying this by the number of protons 3 (4) gives .45 eV (0.5 eV) so that 3 dark protons are needed and 1 goes to ADP to give ADP. This gives a nice fit in both cases.

2. It is claimed that the free energy ΔG liberated in ATP→ ADP is the same as in ADP→ AMP. If ΔS matters, one has for the liberated free energy - metabolic energy currency -

$$\Delta G = \Delta E + T\Delta S \ . \quad (3.8)$$

$\Delta G = -.5eV < 0$, the nominal value of metabolic energy currency, holds true approximately.

3.2.3 Can the free energies liberated in ATP→ ADP and ADP → AMP be the same?

The condition that the metabolic energies as free energy changes are same for various options gives the following conditions.

1. For open and closet flux tube option one would obtain the condition:

$$\begin{aligned} Open \quad & \tfrac{3}{2}E_0 + E_s + [K(3) - K(2)]E_P + T\Delta S(ATP \to ADP) = \\ & E_0 + E_s + [K(2) - K(1)]E_P + T\Delta S(ADP \to AMP) \\ Closed \quad & 2E_0 + 2E_s + [K(3) - K(2)]E_P + T\Delta S(ATP \to ADP) = \\ & E_0 + E_s + [K(2) - K(1)]E_P + T\Delta S(ADP \to AMP) \ . \end{aligned} \quad (3.9)$$

We obtain the following results form $K = N^2$ and $K = N$ options respectively:

$$\begin{aligned} (Open, K = N^2): & \quad E_0 = -10E_P - 2X \ , \\ (Open, K = N): & \quad E_0 = -2X \ , \\ (Closed, K = N^2): & \quad E_0 + E_s = -2E_P - X \ , \\ (Closed, K = N): & \quad E_0 + E_s = -X \ , \end{aligned}$$

$$X = \Delta S(ATP \to ADP) - \Delta S(ADP \to AMP) \ . \quad (3.10)$$

For $K = N^2$ option E_0 is positive even when $X = 0$ is true. For $K = N$ $E_0 = 0$ holds true for $X = 0$ and one must have $X < 0$ meaning that the entropy increase in ADP$\to AMP$ is larger than in ATP$\to ADP$.

2. One obtains the following values for E_0 in various cases. All terms are manifestly positive in the expressions as they should be.

$$\begin{aligned}(Open, K = N^2): &\quad E_0 = -10E_P - 2X \ ,\\ (Open, K = N): &\quad E_0 = -2X \ ,\\ (Closed, K = N^2): &\quad E_0 = -2E_P - X - E_s \ ,\\ (Closed, K = N): &\quad E_0 = -X - E_s \ ,\end{aligned}$$

$$X = \Delta S(ATP \to ADP) - \Delta S(ADP \to AMP) \ . \tag{3.11}$$

3. Liberated free energy can be positive in all cases unless $E_P < 0$ has too large a magnitude.

$$\begin{aligned}(Open, K = N^2): &\quad \Delta G = -7E_P - 2X + Y \ ,\\ (Open, K = N): &\quad \Delta G = E_P - 2X + Y \ ,\\ (Closed, K = N^2): &\quad \Delta G = -E_P - X + Y \ ,\\ (Closed, K = N): &\quad \Delta G = E_P - X + Y \ ,\end{aligned}$$

$$X = -T(\Delta S(ATP \to ADP) - \Delta S(ADP \to AMP)) > 0 \ ,$$
$$Y = T\Delta S(ADP \to AMP) > 0 \ .$$

One can argue that $\Delta S > 0$ in both reactions since the number of $h_{eff}/h_0 > 0$ protons decreases and the "IQ" of the system decreases. Hence one has $Y > 0$. The term E_P in $K = N$ case is negative. The term proportional to X is positive for all cases if $X < 0$ is true. This would mean that $\Delta S(ATP \to ADP) < \Delta S(ADP \to AMP)$. Entropy would increase more in the the latter reaction. $K = N^2$ options are favored and the most favored is (Open, $K = N^2$) option: open flux tube with 3 dark protons interacting with phosphate charges like single charge of 3 units as the identification as a Galois confined state suggests.

4 Conclusions

The number theoretic vision about TGD predicts that dark matter identified as phases of ordinary matter with effective Planck constant $h_{eff} > h$ residing at MB is a key player of physics in all length scales, and especially so in quantum biology and even ordinary chemistry.

The key implication would be quantum coherence at MB in all scales and MB serving as the master of ordinary matter would induce this coherence to the level of ordinary matter. This would explain the coherence of living matter remaining a mystery in the approach to biology as nothing but chemistry.

Second key implication would also involve ZEO: quantum self-organization would rely on time reversal occuring in the ZEO counterparts of ordinary state function reductions. Time reversal would make possible self-organized quantum criticality explaining the paradoxical ability of living systems to stay near criticality: repulsion in the reverse time direction looks like attraction in the opposite time direction.

Concerning chemistry and biochemistry, the first prediction is that valence electrons can be dark. This would explain why valence bond energies increase along the rows of the periodic table, and why the molecules involving elements at the right end of the table serve as carriers of metabolic energy. Oxidation and reduction could be seen as transfer of dark valence electrons. Also dark protons are possible. Hydrogen bonds between say water molecules could have dark variants and induce long range quantum correlations explaining the numerous thermo-dynamical anomalies of water as reflecting the neglect of the dark matter in the modelling of water.

Pollack effect would have interpretation as a phase transition in which external energy feed drives protons from exclusion zones (EZs) to dark protons at flux tubes of MB and leads to charge separation. This would explain the negative charge of the cell. Dark protons would stabilize DNA by screening its negative charge of DNA. Dark proton triplets would give rise to dark proton analogs of DNA, RNA, tRNA, and amino acids (briefly information molecules) and realize genetic code in the sense that there is a natural correspondence with DNA and amino-acids for which the number of DNAs coding for given amino acid are the same as for the vertebrate code. Ordinary chemical realization of the code would be kind of mimicry.

TGD predicts also a second realization of genetic code as dark photon triplets making possible communications between dark variants of information molecules by frequency and energy resonance. The transformation of dark photons to bio-photons would also allow coupling to the information molecules by energy resonance. Energy resonance occurring at cyclotron energies of dark particles could explain the selection of the basic biomolecules.

Received February 23, 2021; Accepted October 1, 2021

References

[1] Icosahedron. Wikipedia article. Available at: http://en.wikipedia.org/wiki/icosahedron.

[2] Tetrahedron. Wikipedia article. Available at: http://en.wikipedia.org/wiki/tetrahedron.

[3] Li C. Optical Four-Wave Coupling Process. 2017. In: Nonlinear Optics. Springer, Singapore. https://doi.org/10.1007/978-981-10-1488-8_4.

[4] Chatterjee S et al. Lifshitz transition from valence fluctuations in YbAl3. *Nature Communications*, 8(852), 2017. Available at: https://www.nature.com/articles/s41467-017-00946-1.

[5] Lin X et al. Beating the thermodynamic limit with photo-activation of n-doping in organic semiconductors. *Nature Materials*, 16:1209–1215, 2017. Available at: https://www.nature.com/articles/nmat5027.

[6] Mills R et al. Spectroscopic and NMR identification of novel hybrid ions in fractional quantum energy states formed by an exothermic reaction of atomic hydrogen with certain catalysts, 2003. Available at: http://www.blacklightpower.com/techpapers.html.

[7] Nottale L Da Rocha D. Gravitational Structure Formation in Scale Relativity, 2003. Available at: http://arxiv.org/abs/astro-ph/0310036.

[8] ATP hydrolysis. Wikipedia article. Available at: http://en.wikipedia.org/wiki/ATP_hydrolysis.

[9] Microtubule. Wikipedia article. Available at: http://en.wikipedia.org/wiki/Microtubule.

[10] The Fourth Phase of Water: Dr. Gerald Pollack at TEDxGuelphU, 2014. Available at: https://www.youtube.com/watch?v=i-T7tCMUDXU.

[11] Smith C. *Learning From Water, A Possible Quantum Computing Medium*. CHAOS, 2001.

[12] Bonora M et al. ATP synthesis and storage. *Purinergic Signal*, 8(3):343–357, 2012. Available at: https://pubmed.ncbi.nlm.nih.gov/22528680/.

[13] Ling GN et al. Experimental confirmation, from model studies, of a key prediction of the polarized multilayer theory of cell water. *Physiological Chem & Phys*, 10(1):87–88, 1978.

[14] Perry et al. Mitochondrial membrane potential probes and the proton gradient: a practical usage guide. *BioTechniques*, 50(2), 2018. Available at: https://www.future-science.com/doi/10.2144/000113610.

[15] Purves D et al. *What are basal ganglia?* 2001. August. Available at: https://www.neuroscientificallychallenged.com/blog/what-are-basal-ganglia.

[16] Ling G. Three sets of independent disproofs against the membrane-pump theory, 1978. Available at: http://www.gilbertling.org/lp6a.htm.

[17] Ling G. Oxidative phosphorylation and mitochondrial physiology: a critical review of chemiosmotic theory and reinterpretation by the association-induction hypothesis. *Physiol. Chem. & Physics*, 13, 1981.

[18] Ling GN. *A physical theory of the living state: the association-induction hypothesis; with considerations of the mechanics involved in ionic specificity.* Blaisdell Pub. Co., New York, 1962.

[19] Ling GN. Maintenance of low sodium and high potassium levels in resting muscle cells. *J Physiol (Cambridge)*, pages 105–123, 1978.

[20] Nwamba OC. Membranes as the third genetic code. *Mol Biol Rep*, 47(5):4093–4097, 2020. Available at: https://pubmed.ncbi.nlm.nih.gov/32279211/.

[21] Zhao Q Pollack GH, Figueroa X. Molecules, water, and radiant energy: new clues for the origin of life. *Int J Mol Sci*, 10:1419–1429, 2009. Available at: http://tinyurl.com/ntkfhlc.

[22] Blackman CF. *Effect of Electrical and Magnetic Fields on the Nervous System*, pages 331–355. Plenum, New York, 1994.

[23] Hameroff S Graddock T, Tuszynksi A. Cytoskeletal Signaling: Is Memory Encoded in Microtubule Lattices by CaMKII Phosphorylation? *PLoS Comput Biol*, 8(3), 2012. Available at: http://www.ploscompbiol.org/article/info:doi/10.1371/journal.pcbi.1002421.

[24] Pitkänen M. DNA and Water Memory: Comments on Montagnier Group's Recent Findings. *DNA Decipher Journal*, 1(1), 2011. See also http://tgtheory.fi/public_html/articles/mont.pdf.

[25] Pitkänen M. Evolution in Many-Sheeted Space-Time: Big Vision. *DNA Decipher Journal*, 3(2), 2013. See also http://tgtheory.fi/pdfpool/prebio.pdf.

[26] Pitkänen M. Evolution in Many-Sheeted Space-Time: Genetic Code. *DNA Decipher Journal*, 3(2), 2013. See also http://tgtheory.fi/pdfpool/prebio.pdf.

[27] Pitkänen M. Music, Biology and Natural Geometry (Part I). *DNA Decipher Journal*, 4(2), 2014. See also http://tgtheory.fi/public_html/articles/harmonytheory.pdf.

[28] Pitkänen M. Music, Biology and Natural Geometry (Part II). *DNA Decipher Journal*, 4(2), 2014. See also http://tgtheory.fi/public_html/articles/harmonytheory.pdf.

[29] Pitkänen M. Maintenance problem for Earth's magnetic field. Available at: http://tgdtheory.fi/public_html/articles/Bmaintenance.pdf., 2015.

[30] Pitkänen M. Cold Fusion Again. *Pre-Space-Time Journal*, 7(2), 2016. See also http://tgtheory.fi/public_html/articles/cfagain.pdf.

[31] Pitkänen M. Philosophy of Adelic Physics. In *Trends and Mathematical Methods in Interdisciplinary Mathematical Sciences*, pages 241–319. Springer. Available at: https://link.springer.com/chapter/10.1007/978-3-319-55612-3_11, 2017.

[32] Pitkänen M. Is It Cold Fusion, Low Energy Nuclear Reactions or Dark Nuclear Synthesis? *Pre-Space-Time Journal*, 8(13), 2017. See also http://tgtheory.fi/public_html/articles/krivit.pdf.

[33] Pitkänen M. On Hydrinos Again. *Pre-Space-Time Journal*, 8(1), 2017. See also http://tgtheory.fi/public_html/articles/Millsagain.pdf.

[34] Pitkänen M. On the Mysteriously Disappearing Valence Electrons of Rare Earth Metals & Hierarchy of Planck Constants. *Pre-Space-Time Journal*, 8(13), 2017. See also http://tgtheory.fi/public_html/articles/rareearth.pdf.

[35] Pitkänen M. On the Correspondence of Dark Nuclear Genetic Code & Ordinary Genetic Code. *DNA Decipher Journal*, 8(1), 2018. See also http://tgtheory.fi/public_html/articles/codedarkcode.pdf.

[36] Pitkänen M. Emotions & RNA. *DNA Decipher Journal*, 8(2), 2018. See also http://tgtheory.fi/public_html/articles/synapticmoods.pdf.

[37] Pitkänen M. On the Physical Interpretation of the Velocity Parameter in the Formula for Gravitational Planck Constant. *Pre-Space-Time Journal*, 9(7), 2018. See also http://tgtheory.fi/public_html/articles/vzero.pdf.

[38] Pitkänen M. An Overall View about Models of Genetic Code & Bio-harmony. *DNA Decipher Journal*, 9(2), 2019. See also http://tgtheory.fi/public_html/articles/gcharm.pdf.

[39] Pitkänen M. New Aspects of $M^8 - H$ Duality. *Pre-Space-Time Journal*, 10(6), 2019. See also http://tgtheory.fi/public_html/articles/M8Hduality.pdf.

[40] Pitkänen M. Results about Dark DNA & Remote DNA Replication. *DNA Decipher Journal*, 10(1), 2020. See also http://tgtheory.fi/public_html/articles/darkdnanew.pdf.

[41] Pitkänen M. The Dynamics of State Function Reductions as Quantum Measurement Cascades. *Pre-Space-Time Journal*, 11(2), 2020. See also http://tgtheory.fi/public_html/articles/hSSFRGalois.pdf.

Exploration

MeshCODE Theory from TGD Point of View

Matti Pitkänen [1]

Abstract

Benjamin Goult has made an interesting proposal in an article *The Mechanical Basis of Memory the MeshCODE Theory* published in Frontiers of Molecular Neuroscience. The proposal is that the cell - or at least synaptic contacts - realize mechanical computation in terms of adhesive structures consisting of hundreds of proteins known as talins, which act as force sensors. Talins are connected to integrins in the extracellular matrix, to each other, and to the actins in the cell interior. This proposal has far reaching consequences for understanding formation of memomies as behaviors at the synaptic level. This proposal does not conform with the TGD vision but inspires a series of questions leading to a rather detailed general vision for how magnetic body (MB) receives sensory input from biological body (BB) coded into dark 3N-photons a representing genes with N codons and as a response activates corresponding genes, RNA or proteins as a reaction. Sensory input and the response to it would be coded by the same dark genes. Since synaptic adhesion structures relate to memories as behaviors and since the TGD based view about memories distinguishes between memories as behaviours and memories as episodal/sensory memories, a discussion about memories according to TGD is included.

1 Introduction

Benjamin Goult has made an interesting proposal in the article *The Mechanical Basis of Memory the MeshCODE Theory* [9] (https://cutt.ly/WzlrmrM) published in Frontiers of Molecular Neuroscience in 25 February 2021.

The proposal is that the cell or at least synaptic contacts realize mechanical computation in terms of adhesive structures consisting of hundreds of proteins known as talins, which act as force sensors. Talins are connected to integrins in the extracellular matrix, to each other, and to the actins in the cell interior.

This proposal does not conform with the TGD vision but inspires a series of questions leading to a rather detailed general vision for how magnetic body (MB) receives sensory input from biological body (BB) coded into dark 3N-photons representing genes with N codons and as a response activates same but differently realized genes, RNA or corresponding proteins as a reaction [14, 15, 23, 25, 26]. This would mean a universal response function assigning to sensory input a unique response. Sensory input would code the response to it in terms of dark genes, which also generalize in TGD framework.

1.1 Some basic facts

The role of a protein known as talin [7] (https://cutt.ly/OzNTvPf) is the topic of the article. Talin is associated with the cell-substratum contact and mechanically couples cytoskeleton and extracellular matrix (ECM) together. Adhesion units formed by integrin coupling to ECM, talin, and actin at cytoskeleton side form adhesion structures consisting of hundreds of adhesion units.

It is good to begin with by listing some basic definitions and facts.

1. Cytoskeleton [2] (https://cutt.ly/xzNTT8s) consists of microfilaments (actin), intermediate filaments, and microtubules (MTs) which in neurons are called neurotubules. Neurons contain neurotubules (NTs) [6] (https://cutt.ly/BzNTZqY) generated at MT organizing center (MTOC) and transferred to dendrites and axon, where they are parallel to the neuronal surface.

---

[1]Correspondence: Matti Pitkänen http://tgdtheory.com/. Address: Rinnekatu 2-4 A8, 03620, Karkkila, Finland. Email: matpitka6@gmail.com.

The cytoskeleton of an ordinary cell has as basic building bricks MTs and microfilaments and intermediate filaments. Both MTs and NTs are polarized. The + ends of MTs are at MTOC. + ends of NTs point towards the axon terminal and - end to the parent neuron. NTs in dendrites have mixed polarities.

2. ECM [3] (https://cutt.ly/5zNYtP6) is a three-dimensional network consisting of extracellular macromolecules and minerals, such as collagen, enzymes, glycoproteins and hydroxyapatite that provide structural and biochemical support to surrounding cells. Cell adhesion, cell-to-cell communication and differentiation are common functions of the ECM.

3. Integrins [4] (https://cutt.ly/xzNYk7n) are transmembrane receptors that facilitate cell-cell and cell-extracellular matrix (ECM) adhesion. Upon ligand binding, integrins activate signal transduction pathways that mediate cellular signals such as regulation of the cell cycle, organization of the intracellular cytoskeleton, and movement of new receptors to the cell membrane. The presence of integrins allows rapid and flexible responses to events at the cell surface (e.g. signal platelets to initiate an interaction with coagulation factors).

4. Actins [1] (https://cutt.ly/LzNYEo9) are a family of globular multi-functional proteins that form microfilaments. It is found in essentially all eukaryotic cells, where it may be present at a concentration of over 100 μM; its mass is roughly 42-kDa, with a diameter of 4 to 7 nm. An actin protein is the monomeric subunit of two types of filaments in cells: microfilaments, one of the three major components of the cytoskeleton, and thin filaments, part of the contractile apparatus in muscle cells.

One can visualizetalin as a spring between cytoskeleton and ECM. Talincouples directly to integrins at ECM side and either indirectly or directly to actin at cytoskeleton side. Talin's role is to be a rope in a "tug-of-war" between integrins at ECM and actin and it acts as a force sensor and could give rise to a molecular sense of touch based on force.

The part of talin subject to forces from the cellular interior and environment consists of 13 proteins domains which can be in two thermodynamically stable states analogous to the opposite magnetizations of ferromagnet and the domain exhibits hysteresis curve under a varying external force. The phases correspond folded and unfolded configuration looking like a straight bar. The two phases can be labelled by a bit and the proposal is that the talin conformations define 13 bits.

The domains are not identical so that each equilibrium state under varying external net force could correspond to a unique configuration in which domains are folded or unfolded. If so, talin would serve as a 13-bit force sensor of external forces with finite resolution corresponding to 13 octaves in linear scale. It will be found that the response could actually be determined by 6 bits and correspond to genetic codon.

The abstract of [8] summarizes the functions of talin.

"... Talin forms the core of integrin adhesion complexes by linking integrins directly to actin, increasing the affinity of integrin for ligands (integrin activation) and recruiting numerous proteins. It regulates the strength of integrin adhesion, senses matrix rigidity, increases focal adhesion size in response to force and serves as a platform for the building of the adhesion structure. Finally, the mechano-sensitive structure of talin provides a paradigm for how proteins transduce mechanical signals to chemical signals."

It is clear that talin does not look only a passive sensory receptor. That integrins are not necessary for talins to function implies that they have emerged before integrins in the evolution. It is clear that talins are essential aspect of multicellular life.

1.2 Could adhesion structures act as classical computers?

The proposal of the article [9] relies on computationalism and suggests that talin could be more than a sensory receptor and adhesion structures could act as a computer. The structures formed by the adhesion units consisting of integrin-talin-actin triplets would serve as 13-bit units. Adhesion units would perform mechanical computation based on what authors call MESHcode.

One can argue that mechanical computation requires that adhesion units are isolated from the environment during the computation. This is in conflict with the role as force sensors. A weaker proposal would be that computation occurs only in the synaptic contacts which should be isolated during the computation. The same could take place also in the contacts between neurons and glial cells.

Concerning the synaptic level, a more realistic view to my opinion is that learning as a strengthening of the synaptic strengths corresponds to a development of force equilibrium of adhesion units. Learning could be described as the change of the resting states of the talin units and lead to a higher tension and larger number of unfolded protein domains. Nerve pulse patterns could cause temporary changes of this pattern.

2 TGD interpretation of adhesion units as quantal force sensors

In the TGD framework all communications and control in biology should rely on genetic code whose fundamental realization would be at the level of dark proton sequences forming dark nuclei with $h_{eff} = nh_0 > h$ and dark photons.

Dark proton triplets - light 3-chords - would represent the counterparts for dark DNA, RNA, tRNA, and aminoacids and dark photon triplets could represent dark DNA codons [15, 23, 25, 26]. Number theoretic vision [19, 20] leads to a proposal that not only dark 3-photon 3-proton units act as single particle like units but also dark 3N-photons and 3-N protons do so and represent a gene consisting of N codons. Galois confinement would bind the photons and protons to larger particle units analgous to baryons as composites of 3-quarks.

All communications to MB would use dark 3N-photons coupling to corresponding dark 3N-proton by cyclotron resonances [14, 27, 28]. Therefore 3N-photon as a dynamical gene with N codons would define its own address. Frequency modulation of frequencies of 3N-photon would give rise to a sequence of resonance peaks and the continuous signal would be transformed to a signal analogous to nerve pulse sequence and could realize motor action as a response.

2.1 Magnetic body containing dark matter as the master

MB has a hierarchical onion-like structure with levels labelled by the value of $h_{eff} = nh_0$ giving rise to increasing scales. The dark analogs of DNA, RNA, tRNA, and amino-acids define flux tubes accompanying their ordinary variants with codons realized as dark 3-proton units.

In TGD genetic code in terms of 3-chords would be realized in a universal manner for the simplest tesselation of hyperbolic space known as icosa-tedrahedral honeycomb involving icosahedrons and tetrahedrons (also octahedrons are involved but they would be in passive role)[26]. This would suggest that genetic code using dark proton- and dark photon triplets is realized at all layers of MB. Chemical realization would represent the lowest level in the hierarchy.

The layers of MB with increasing value of h_{eff} would define a hierarchy of abstractions. There is evidence for an effective statistically determined hyperbolic geometry [11] in the sense that neurons functionally but not necessarily spatiall near to each other are near to each other in this effective geometry. This hyperbolic geometry would be realized quite concretely at the level of MB [24] for which hyperbolic geometry of proper time constant hyperboloid of the light-cone gives a concrete meaning.

One particular implication could be that sensory receptors of a given structure (say adhesion units of given cell-environment pair) could communicate their sensory data to neighboring icosa-tetrahedral units of the honeycomb of some layer of MB representing the codons of genetic code. The states of the icosahedrons and tetrahedrons of the honeycomb would be dynamical and selected by the 3-chord (actually pair of 3-chord and conjugate) to actualize genetic codon as 3-quark units assignable to the corresponding triangle of icosahedron or tetrahedron.

This would define sensory representation at MB, and the simplest option is that it automatically determines motor response as a sequence of resonance peaks communicated back to the biological body (BB) where they would initiate gene expression, RNA or protein activity, MT activity, or nerve pulse activity. The feedback would be directly to DNA (or RNA, amino-acid of protein, or even tRNA, microbuli, or cell membrane).

The biochemical motor actions of MB would be realized as bursts of dark cyclotron 3N-photons induced by the cyclotron resonances at MB transforming to ordinary photons (biophotons or IR photons with energy above thermal energy) controlling biochemistry by inducing molecular transitions.

This condition constrains the value of h_{eff} for a layer of MB. The size of the layer should be of the order of wavelengths involved. For valence bonds the values of $h_{eff} = h_{em}$ would be rather small and assignable with small layers of MB. For frequencies in EEG range the large value of gravitational Planck constant $h_{eff} = h_{gr}$ [21, 14] assignable to the gravitational flux tubes would guarantee that the energies are in the required range.

The following picture about how sensory input induces gene expression or some other activity with communication and control realized in terms of genetic code might apply completely generally, not only in the case of adhesion units.

1. Suppose the sensory receptors of a given structure (say adhesion units of a given cell) are organized into coherent structures in the sense that the signals from them go along flux tubes to nearby cells of icosa-tetrahedral honeycomb at some layer of MB.

   Adhesion structures consting of few hundred adhesion units are indeed connected to each other. Coherence would be forced by the quantum coherence at the level of MB as a forced coherence. One could assume that the cells of the honeycombinvolved are organized linearly but even 2-D and 3-D structures are possible.

   For a structure consisting of N units, the dark 3N-photon signal would define a dark gene of N codons. The nice feature of the representation is that there is no need to organize the sensory receptors (say adhesion units) linearly at the level of the cell. The level of ordinary biomatter would be like RAM with ordering realized at the level of MB.

2. The naive picture is that if the dynamical gene realized in this manner has a dark counterpart at the level of flux tube accompanying DNA, gene expression could be initiated automatically as a feedback signal realized as a sequence of resonance peaks. Also RNA, proteins or MTs could be activated in an analogous manner.

   There would be a one-one correspondence between sensory inputs to MB and corresponding gene expressions and give a meaning for the genetic code. All sensory inputs to MB would be realized as N-genes in terms of generalized Josephson radiation which is frequency modulated and generates a sequence of resonance peaks inducing gene expression or RNA and protein activation.

3. The dynamical gene at MB need not correspond to an existing or expressible gene so that the response is not possible. This would give rise to an evolutionary pressure. Epigenesis controlled by MB could make the gene expressible. Also a suitable mutation for existing gene or emergence of new gene could produce the needed gene. Whether MB is able to induce this kind of mutations is an interesting question. Could a dark gene as a flux tube containing dark proton sequence representing the desired gene pair with ordinary DNA codons and give rise to a new gene?

   Or could MB "use scissors" to replace codon-anticodon pairs in an existing gene: this would mean reconnection of a closed flux tube pair containing the codon-anticodon pairs of the added gene fragment. Could a piece of dark DNA as a flux tube carrying the dark proton sequence pair with ordinary DNA codons and give rise to a new gene? Or could one add to an existing gene a piece represented as a dark DNA paired with the ordinary DNA. Most viruses have single stranded RNA genomes. Bacteriophages have double stranded DNA genomes. They are known to give rise to the modifications of the genome. Could these DNA modifications be induced by a reconnection of darkmagnetic flux tubes.

2.2 Universality of the genetic code and its higher dimensional representations

If genetic code at space-time surface is induced from a universal code assignable to the icosa-tetrahedral honeycomb of hyperbolic 3-space, representations of genetic code with dimensions $D = 0, 1, 2, 3$ are possible as induced representations. The codons associated with the cells of honeycombes projected to the space-time surface would define the induced codons [26].

tRNA would be a 0-D representation and DNA, RNA, amino acids would be 1-D representations of the code. Also higher-dimensional representations are possible and could be associated with the basic biological structures.

1. I have proposed that cell membrane defines a 2-D representation of the genetic code [26]. Also microtubuli could define a 2-D representation of genetic code. These 2-D representation could be dynamical and independent of genome and make genome dynamical. This would be a biological analog for AI able to write genes as program modules needed in a given situation.

2. Could a 3-D representation of genetic code be associated with the ECM and make it possible for MB to receive sensory input from ECM and control it? This layer of MB could also receive sensory information also from adhesive structures. The frequency range involved would be probably below EEG frequencies or at least below conscious frequencies since we do not experience the interior of body consciously and the time scale of dynamics is slow as compared to EEG scales.

   Hydrozyapatite molecules are present in bones forming a part of ECM. Fisher has proposed that the Posner molecules associated with hydroxyapatite molecules could have important role in quantum biology [12]. This inspired the proposal that they provide a realization of genetic code [16]. One cannot exclude the possibility that the code is 3-D. This would fit with the general idea that the genetic code serves as a universal code for communications and control.

2.3 Some TGD inspired numerology

If one takes the proposed general picture seriously, one must ask how the 13-bits codons assignable to talins and MTs could reduce to genetic codons. It is good to start with numerology or should one call it physics inspired poor man's number theory.

1. The number of protein domains in talin is 13. Also the number of tubulin dimers in 13-tubulin unit of MT/neurotubule appearing in cytoskeleton is 13. Could one think of communication between MTs and talins using 13 bit code? Or could the code using 13 bits be for some reason special? Could this code somehow reduce to the proposed universal 6-bit code defined by genetic code?

2. There are 4 protein domains consisting of 4 alpha helices and 9 domains with 5 alpha helices. This gives 61 alpha helices altogether. Numerologist might notice that 61 is the number of DNA codons with stop codons excluded. Could one assign to helices genetic codons and could these configurations labelled by 61 bits code for genes with length not longer than 61 units?

3. Numerologist might also notice that both $M_{13} = 2^{13} - 1$ and $M^{61} = 2^{61} - 1$ are Mersenne primes. If one has n bits and does not count the configuration with all bits 0 but assuming that at least single bit is always equal to 1, one has 2^{n-1} full bits.

   For M_{13} this corresponds to 12 full bits which corresponds to 2 genetic codons. To obtain 2 codons, single fixed talin should be unfolded and represent 1. Could this have interpretation in terms of a force threshold? One can argue that there is some minimal force unfolding some fixed talin. If the force is below the threshold, there is no need to communicate. Also in the case of MT the conformation of preferred tubulin, say the first or last one in 13-unit should always correspond to 1.

4. One cannot exclude the possibility that the responses of talin units correspond to two independent codons. This could be true also for 13-bit units MTs.

   The alternative option is that both talins and 13-tubulin units of MT correspond to codon-anticodon pairs so that information content would reduce to that of single DNA codon. Half of the bits would serve as check bits. Also the purpose of the conjugate strand of DNA would be to serve as check codons.

   If this is the case, the adhesion unit would have only 2^6 different responses and would represent a genetic codon. The number of talins is few hundred that this would correspond to a DNA sequence of length of order 10^{-7} meters. In the case of MT 6 bits would be check bits.

5. The proposal would have far reaching consequences: the genetic code realized by MTs and talins would be dynamical rather than fixed and could represent a step to a higher evolutionary level.

6. The dynamics of the codon or of a pair of pair of independent codons assignable to the adhesion unit would mean change of the "sensory codon" possibly corresponding to a real codon assignable to it. The slow time variation of the gene assignable to the collection of adhesion units could define varying gene expression or some other activations (of say microtubuline).

These speculations encourage the question whether the codon-anticodon pairs possibly assignable to adhesion units integrate to sequences or perhaps even 2-D structures representing 2-D adhesion structures of DNA codon-anticodon pairs defining genes.

If these 2-D honeycomb structures at the level of MB decompose to piles of 1-D structures as microtubules do, they could even induce the expression of gene groups. Also 2-D gene expression in terms of microtubules modifying the cytoskeleton can be considered. Note that the honeycomb structures are not needed at the level of ordinary biomatter.

2.4 A simple model for the adhesion units

In TGD framework magnetic body (MB) containing dark matter controls ordinary living matter. MB receives sensory input from organism in terms of dark Josephson radiation arriving from cell membranes acting as generalized Josephson juctions. Sensory information is coded by the modulation of membrane potential. For ordinary cells only small modulations of membrane potential would induce modulations of Josephson frequency. For neurons nerve pulse patterns introduce more drastic modulation.

1. The two states of the protein domains could correspond to different values of h_{eff}. The reduction of h_{eff} at the magnetic flux tube accompanying the protein would induce the shortening of the flux tube associated with the unfolded protein to the folded configuration.

2. Cohesion units would aserve as sources of sensory information about the net force acting on the cohesion unit and coded by 13 bits unless the bits are independent. For instance, different bits would correspond to different signals, say different frequencies of dark photons. If one takes the interpretation as a pair of codons seriously, the signal could consist of a dark 3-chord and its conjugate 3-chord sent to MB and defining at the MB a representation of gene to be possibly activated.

3. Josephson radiation as dark 3-photons from the part of the cell membrane considered would mediate the 13 bit signal defined coded to a local change of membrane potential with 2^{12} values defining 12 octaves if there is threshold corresponding to activation of a preferred talin. Note that the frequencies audible for humans are in the range 20 Hz- 20 kHz and correspond to 10 octaves.

4. MB would receive the sensory input and react by possibly sending control signal to DNA inducing gene expression or inducing activity of proteins or RNA. This means that talin molecules would not be active but MB receiving the sensory input from adhesion units.

MB could also send control signal to microtubuli if MT contains a sequence of 13-tubulin units corresponding to the dynamical gene [5] (https://cutt.ly/MzNYBsZ). This would reflect itself in the dynamics of MTs. This control loop would modify the force equilibrium by a modification of the shape of the cell.

5. MTs could represent an evolutionary step making the genome dynamical and independent of genes and extending ordinary genome as the microtubular response possible for eukariotes suggests. Also the long MTs inside axons conform with this interpretation.

6. MTs are highly dynamical. Their lengths are continually varying. According to "search-and-catch" model MTs inside cells are scanning their 3-D environment and whey the find a target attach to it and MT is stabilized. This conforms with general vision about U-shaped dynamical flux tubes serving as tentables and forming a reconnection with a similar U-tube of the target. Immune system would be rely on this mechanism at the fundamental level and allow the system to detect and catch invader molecules on basis of their cyclotron energy/frequency spectrum [13, 14].

7. The general vision suggests that the feedback loop should involve also microfilaments and intermediate filaments. It would be interesting to see whether the the structure of microfilaments and intermediate filaments could allow realization of the counterpart of genetic code. The basic signature are GTP and ATP molecules providing metabolic energy for motor action.

3 An application to memory and learning

Since the increase of synaptic strengths is believed to be behind the formation of memories as behaviors and habits, it is appropriate to discuss the notion of memory in TGD framework and consider connections with the model for the adhesion units at synaptic contacts.

The major issue with memory is potentiation (repeat of same memory which facilitatesmemory recall and learning) and amnesia, Alzheimer disease and memory when dreaming. There should be a compatible explanation for these phenomena.

In TGD one distinguishes between two kinds of memories. Episodal-/sensory memories and memories as associations/learned behaviors.

3.1 Memories as learned behaviors

Neuroscience explains learned behaviors in terms of strengthening of synaptic contacts and I believe that this is part of the story.

The formation of associations in conditioning is a highly emotional process and here the surprising finding [10] (see http://tinyurl.com/ycqxyeqk) few years ago (roughly) was helpful. The popular article "Scientists Sucked a Memory Out of a Snail and Stuck It in Another Snail" tells about the finding (see http://tinyurl.com/y92w39gs).

The RNA of a sea snail which had learned by (presumably painful) stimulus a behavior was scattered on the neuronal tissue of another sea snail in a Petri dish. The neuronal tissue learned the same behavior!

The TGD based explanation is following.

1. Emotions are realized already at the molecular level [22] in terms of music of light - bioharmony [15, 23, 25, 26]. The emotional stimulus at the MB of RNA induced learning by changing the allowed 3-chords of bioharmony. Also the sequences of 3-chords characterizing 3N-genes and other basic linear biomolecules changed. The resonant couplings to the basic biomolecules changed so that also chemical behavior changed.

2. The emotional state of the conditioned seanail RNA infected the RNAs and probably also DNAs and proteins of neurons and induced learning.

3. Synaptic strengths had to change and the molecular emotions as music of light would have induced this.

If the idea about mechanical control of synaptic strengths by talin molecules by push and pull from ECM and cytoskeleton is correct, the molecular mood had to induce a strong force changing the talin conformations. Emotion would quite concretely correspond to a force!

This would have induced a reaction at the level of microtubules with the mediary of MB as a response making the change permanent. Neurotubules of the cytoskeleton in dendrites and axons would be involved in realizing the learning as a permanent change.

3.2 Potentiation and two kinds of memories

The notion of potentiation applies to both kinds of memories.

1. The repetition of stimulus generating the learned behavior increases the synaptic strength. Perhapsby inducing a memory recall of the emotional experience at molecular level.

2. Potentiation for sensory memories creates an almost copy of sensory memory mental image at "geometric now": the re-experience and the more one has these almost copies in the geometric future of "geometric now", the higher the probability that the attempt to remember by sending dark photon signals to the future hits the memory mental image are successful. The latest memory recalls create memories mental images nearest to "geometric now" and the probability for memory recall is highest for them.

Why oldest sensory memories are those which survive when one begins to lose memories at old age?

1. There are a lot of almost copies about the oldest memories: does this mean that the memory recall has a higher probability to be successful?

2. One can also argue that the memory mental images of young age have also gone through a long sequenceof re-incarnations which have gradually increased the value of h_{eff}.

Large h_{eff} means that the frequency f needed to produce a dark photon with energy $E = h_{eff}f$ in biophoton range is lower and therefore the period $T = 1/f$ is longer. Uncertainty Principle says that the time period over which memories are optimally recalled is of order $T = 1/f$.

3.3 Amnesia, Alzheimer, and why we forget dreams so fast

Amnesia might relate to the inability to recall sensory memories by sending signals with a correct frequency to the memory mental images. The energy of the dark photons is proportional to h_{eff} and if it is reduced in the recalling end as tends to happen in the absence of metabolic energy feed, the ability to recall memories is weakened or lost. For instance, alcoholism can lead to a loss of memory recall and this could be the reason.

Alzheimer means a loss of memories as behaviors and inability go generate new ones. In TGD framework [18] the weakening of the synaptic connections would make the build up of connection between magnetic flux tubes associated with presynaptic dendrite and postsynaptic axon and the dark photon signal could not propagate because the connection is broken.

Also the propagation along axonal flux tubes could be impossible or highly attenuated if the value of h_{eff} for them is reduced. Also the energy for a given freqeuency would be reduced below the biophoton energy range.

Why do we forget dreams so fast? We do not remember anything about sleep without dreams. In ZEO this can be understood if sleep corresponds to "small death" for an appropriate layer of MB meaning re-incarnation with an opposite arrow of time. Dreams would correspond to states in which part of the brain is awake and possibly receives information from the sleeping part of the brain realized as a dream. Dream would be due to a communication of virtual sensory input from MB with opposite arrow of time to sensory organs.

This does not yet explain why we forget dreams so fast. As the memory image ages, it shifts to the future of "geometric now" in CD, and the needed frequency as inverse of the age decreases. Could it be that we cannot generate the frequencies of dark photons needed for the memory recall.

3.4 Memories change

Episodal memories are not carved in stone. They are modified in memory recalls. In TGD framework, the modification of (episodal) memory mental images is unavoidable. Memory mental images are living entities and evolve re-incarnation by re-incarnation. Memory recalls are basically analogous to quantum measurements of memory mental images induced BSFR and quantum measurement indeed changes the state of the system measured.

1. The sub-selves of self as mental images continue to live at sub-CDs which in the proposed model drift to the geometric future of CD increasing SSFR by SSFR. These sub-CDs experience BSFRs and evolve incarnation by incarnation. In general evolution happens and they become smarter and wiser. Memories are indeed said to grow sweeter in time.

2. Each memory recall must take the memory subself to a state in which it has arrow of time opposite to that of recaller so that the signal about the memory propagates to the geometric past to "geometric now" [the ball at center of CD at which future and past directed cones glued together].

   The BSFR for memory subself with the same arrow of time as recaller induces memory recall. Memory recall is a murderous process. If the memory recall occurs spontaneously, the murder is not not the recaller.

3.5 Confabulation

The phenomenon of confabulation relates most probably to episodal/sensory memories, not memories as behaviors and habits. Confabulation could be understood in the following manner. Memory mental images are just glimpses about what happened since only those aspects of the event which receive the attention form memory mental images. Memory recaller builds a logical sounding story around these glimpses so that confabulation is unavoidable.

Even our sensory perception is fabrication of stories [17]. Sensory organs are seats of primary sensory experience and there is feedback from MB and brain to sensory organs as virtual input. This feedback loop generates standardized mental images by pattern completions and recognition.

If the sensory input is meager the story can be non-realistic as I know as a person with a poor eye sight. REM dreams and hallucinations are an excellent example of this: in this case there is only virtual sensory input present.

Acknowledgements: I am grateful for Reza Rastmanesh for the questions about memory that inspired the last section of the article.

Received March 28; Accepted October 1, 2021

References

[1] Actin. Available at: http://en.wikipedia.org/wiki/Actin.

[2] Cytoskeleton. Available at: http://en.wikipedia.org/wiki/Cytoskeleton.

[3] Extracellular matrix. Available at: http://en.wikipedia.org/wiki/Extracellular_matrix.

[4] Integrin. Available at: http://en.wikipedia.org/wiki/Integrin.

[5] Microtubule. Wikipedia article. Available at: http://en.wikipedia.org/wiki/Microtubule.

[6] Neurotubule. Available at: http://en.wikipedia.org/wiki/Neurotubule.

[7] Talin. Available at: http://en.wikipedia.org/wiki/Talin.

[8] Brown NH Klapholz B. Talin – the master of integrin adhesions. *Journal of Cell Science*, 130:2435–2446, 2017. Available at: https://jcs.biologists.org/content/130/15/2435.

[9] Goult B. The Mechanical Basis of Memory – the MeshCODE Theory. *Front. Mol. Neurosci.*, 2021. Available at: https://doi.org/10.3389/fnmol.2021.592951.

[10] Bedecarrats A et al. RNA from Trained Aplysia Can Induce an Epigenetic Engram for Long-Term Sensitization in Untrained Aplysia. *eNeuro.0038-18.2018*, 2018. Available at:http://www.eneuro.org/content/early/2018/05/14/ENEURO.0038-18.2018.

[11] Cacciola A et al. Coalescent embedding in the hyperbolic space unsupervisedly discloses the hidden geometry of the brain, 2017. Available at:https://arxiv.org/pdf/1705.04192.pdf.

[12] Fisher MPA. Quantum Cognition: The possibility of processing with nuclear spins in the brain), 2015. Available at: https://arxiv.org/abs/1508.05929.

[13] Pitkänen M. Homeopathy in Many-Sheeted Space-Time. In *Bio-Systems as Conscious Holograms*. Available at: http://tgdtheory.fi/pdfpool/homeoc.pdf, 2006.

[14] Pitkänen M. and Rastmanesh R. The based view about dark matter at the level of molecular biology. In *Genes and Memes: Part II*. Available at: http://tgdtheory.fi/pdfpool/darkchemi.pdf, 2020.

[15] Pitkänen M. Geometric theory of harmony. Available at: http://tgdtheory.fi/public_html/articles/harmonytheory.pdf., 2014.

[16] Pitkänen M. Are lithium, phosphate, and Posner molecule fundamental for quantum biology? Available at: http://tgdtheory.fi/public_html/articles/fisherP.pdf., 2016.

[17] Pitkänen M. DMT, pineal gland, and the new view about sensory perception. Available at: http://tgdtheory.fi/public_html/articles/dmtpineal.pdf., 2017.

[18] Pitkänen M. Is it possible to reverse Alzheimer's disease? Available at: http://tgdtheory.fi/public_html/articles/Alzheimer.pdf., 2017.

[19] Pitkänen M. Philosophy of Adelic Physics. In *Trends and Mathematical Methods in Interdisciplinary Mathematical Sciences*, pages 241–319. Springer.Available at: https://link.springer.com/chapter/10.1007/978-3-319-55612-3_11, 2017.

[20] Pitkänen M. Philosophy of Adelic Physics. Available at: http://tgdtheory.fi/public_html/articles/adelephysics.pdf., 2017.

[21] Pitkänen M. About the physical interpretation of the velocity parameter in the formula for the gravitational Planck constant . Available at: http://tgdtheory.fi/public_html/articles/vzero.pdf., 2018.

[22] Pitkänen M. Emotions as sensory percepts about the state of magnetic body? Available at: http://tgdtheory.fi/public_html/articles/emotions.pdf., 2018.

[23] Pitkänen M. An overall view about models of genetic code and bio-harmony. Available at: `http://tgdtheory.fi/public_html/articles/gcharm.pdf.`, 2019.

[24] Pitkänen M. Could brain be represented as a hyperbolic geometry? Available at: `http://tgdtheory.fi/public_html/articles/hyperbolicbrain.pdf.`, 2020.

[25] Pitkänen M. How to compose beautiful music of light in bio-harmony? `https://tgdtheory.fi/public_html/articles/bioharmony2020.pdf.`, 2020.

[26] Pitkänen M. Is genetic code part of fundamental physics in TGD framework? Available at: `https://tgdtheory.fi/public_html/articles/TIH.pdf.`, 2021.

[27] Pitkänen M and Rastmanesh R. New Physics View about Language: part I. Available at: `http://tgdtheory.fi/public_html/articles/languageTGD1.pdf.`, 2020.

[28] Pitkänen M and Rastmanesh R. New Physics View about Language: part II. Available at: `http://tgdtheory.fi/public_html/articles/languageTGD2.pdf.`, 2020.

In Memoriam

Peter P. Gariaev (1942 - 2020): Discoverer of Phantom DNA Effect & Founder of "Wave Genetics"

Huping Hu* & Maoxin Wu

ABSTRACT

Peter P. Gariaev (1942 - 2020) was the discoverer of phantom DNA effect and founder of "wave genetics". He was a member of DNADJ Advisory Board and published visionary and groundbreaking scientific papers in this journal. He was nominated for Nobel Prize in Medicine for 2021 according to his website and his legacy in science will live on.

Keywords: Peter Gariaev, phantom DNA effect, wave genetics, scientist, visionary, legacy.

We learned earlier this year about the passing of Dr. Peter P. Gariaev in November, 2020 from the Internet [1] and private communications as his passing was not announced on his website [2]. The fields of genetics and medicine have lost a pioneer and visionary and we have lost a dear colleauge and important contributor to this journal.

Peter P. Gariaev (1942 - 2020) was the discoverer of phantom DNA effect and founder of "wave genetics". He was a member of DNADJ Advisory Board and published visionary and groundbreaking scientific papers in this journal. He was nominated for Nobel Prize in Medicine for 2021 according to his website [2].

His contributions to the advancement of science and technology with his colleagues and collaborators include the following [3]:

> 1984 - 1985 - Using correlation method of laser spectroscopy, he revealed the phenomenon of abnormally long damped oscillations DNA gels with specific traits related phenomena return Fermi-Pasta-Ulam (FPU). This can be interpreted as evidence of spontaneous soliton excitation of DNA with the DNA of the new memory elements ("return") type. At the same time by the same method, he discovered the effect of DNA phantom memory that had not been previously well known.

Correspondence: Huping Hu, Ph.D., J.D., QuantumDream Inc., P. O. Box 267, Stony Brook,, NY 11790. E-mail: editor@dnadecipher.com

1992 - Using the method of laser spectroscopy correlation, he discovered the phenomenon of distant instrument of influence on the vibrational dynamics of gels of DNA.

1996 - In conjunction with a corporation, he and his colleagues created bio-radio-electronic and bio-optical systems that simulate some aspects of information-wave processes of the genetic apparatus. These systems combine functional nonliving (opto-electronics engineering) and live (live cells, tissues, organs, metabolic cell-free system), and preparative isolation and/or artificially synthesized genetic structures (chromosomes, DNA, RNA, proteins). Biological substrates used, functionality combined with fiber-avionics are memory elements and the basis of the simplest biocomputer, which is able to control the wave through the defined areas of genetic and metabolic information of biological systems.

1996 - Together with LPI he and his colleagues used the two-photon excitation of an artificial laser-like radiation DNA and chromosomes (superluminescence) as an analogue of the natural emission of photons genetic structures.

2000 – He laid the theoretical and experimental basis for a fundamentally new PLR-spectroscopy (polarization of laser-radio wave spectroscopy) with transition localized photons in radio-wave radiation of any object, including chromosomes, living cells, tissues, and metabolites.

1993, 2000, 2002 - In exploratory experiments, radiation damaged chromosomes of wheat and barley, as well as some "lively" radiation-damaged seeds Arabidopsis thaliana, collected in the Chernobyl nuclear power plant in 1987, were restored.

1999, 2002 – He established a theoretical basis, and made a soft reversible wave introduction of genetic information from DNA into the genome of the animal origin of potatoes and obtained in the 1st and the 2nd generation of a unique plant-animal "hybrid" with signs of unusual morphogenetic that have been lost (not inherited) in the 3rd generation.

2001 - 2002 – He transferred wave genetic information over a distance of about 5 km from the DNA sample extracted from the plant Arabidopsis thaliana line, the plant Arabidopsis thaliana other nearby lines.

2002 - He successfully transferred by a wave of genetic and metabolic information from the pancreas and spleen healthy newborn rats on adult rats suffering from artificially induced diabetes causing the symptoms of diabetes to disappear in a few days.

In additions, many of his important and more recent research results were published in this journal [*e.g.*, 4-6].

We have no doubt that he had made visionary and groundbreaking contributions to the advancement of science and technology. We at DNADJ celebrate his life with our dear readers and thank him for his important contirubtions to this journal – May his legacy in science live on for a long time to come!

References

1. https://www.youtube.com/watch?v=4iNP-cljvUo

2. https://wavegenetics.org/en/

3. Gariaev, P. P., & Leonova, E. A. (2014), The Strange World of Wave Genetics, *DNA Decipher Journal*, 4(1): pp. 39-56.

4. Gariaev, P. P., *et. al.* (2014), Materialization of DNA Fragment in Water through Modulated Electromagnetic Irradiation, *DNA Decipher Journal*, 4(1): pp. 01-02.

5. Gariaev, P. P., Vladychenskaya, I. P., & Leonova, E. A. (2016), PCR Amplification of Phantom DNA Recorded as Potential Quantum Equivalent of Material DNA, *DNA Decipher Journal*, 6(1): pp. 01-11.

6. Gariaev, P. P., Vlasov, P. P., Poltavtseva, R. A., Voloshin, L. L., & Leonova-Gariaeva, E. A. (2018), A Case Report of Tooth Regeneration in Dog through Wave Genetics, *DNA Decipher Journal*, 8(3): pp. 162-165.

In Memoriam

Iona Miller (1949 - 2021): Multitalented Writer, Artist & Visionary

Huping Hu* & Maoxin Wu

ABSTRACT

Iona Miller (1949 - 2021) was a multitalented writer, artist and visionary. She was an inspirational and compassionate human being and a member of the Advisory Board of this journal. She also published scientific papers here. She will be missed by us, our readers/patrons and many others who knew her – Good journey on the other side, Iona, and may your legacy live on!

Keywords: Iona Miller, artist, writer, visionary, nonfiction, multimedia, legacy, compassionate, inspirational.

Iona Miller journeyed to the other side quite unexpectedly on March 26, 2021 [1]. She last communicated on March 4, 2021 with the first author on a scientific topic through Facebook Messinger. She was "a nonfiction writer for the academic and popular press, clinical hypnotherapist (ACHE) and multimedia artist[;] [h]er work is an omni-sensory fusion of intelligence, science-art, new physics, symbolism, source mysticism, futuring, and emergent paradigm shift, creating a unique viewpoint[;] [she was] interested in extraordinary human potential and experience, and the EFFECTS of doctrines of religion, science, psychology, and the arts[;] [she served] on the Advisory Boards of Journal of Consciousness Exploration & Research, DNA Decipher Journal, and Scientific God Journal...." [2].

She was an inspirational and compassionate human being and a truth seeker. She pondered and explored the meanings of life and death through numerous writings and artworks [2].

In an essay/statement entitled "The Mask of Eternity: The Quest for Immortality and the Afterlife" and published in a Special Issue of JCER "Theories of Consciousness and Death" edited by Gregory M. Nixon, Ph.D., she shared the following with the readers [3]:

Correspondence: Huping Hu, Ph.D., J.D., QuantumDream Inc., P. O. Box 267, Stony Brook,, NY 11790. E-mail: editor@dnadecipher.com

When we are gone, only the ultimate question remains. Evidence that consciousness survives death remains elusive. With or without warm, welcoming smiles from relatives we may have loathed in life, it remains our obsession to know what happens when our screen-reality stops, and fades to black. Conscious immortality remains questionable. This writer remains firmly agnostic but enjoys entertaining wishful thinking.

Death is the greatest mystery of life. Buddha rejected the question as useless, according to Jung. Throughout history, it remains a source of wonder, fear, hopefulness, and puzzlement. We seek compassionate ways of dealing with this uncertainty that no discussion of entanglement or holographic memory can assuage.

There is little wonder we tend to fall back on traditional attitudes informed by simplicity, meaningful ceremony, and acceptance. It is something we cannot grasp at all, despite our conceptions of time and space and what might lie beyond them, even if some of our psychic experience seems unbound by spacetime. There is NoWhere to go and we are all going to get there.

We at DNADJ celebrate her life and thank her for her advisory services to the journal. She will be missed by us, our readers/patrons and many others who knew her – Good journey on the other side, Iona, and may your legacy live on!

References

1. https://www.facebook.com/iona.miller
2. https://ionamiller.weebly.com/
3. Miller, I. (2016), The Mask of Eternity: The Quest for Immortality and the Afterlife, *Journal of Consciousness Exploration & Research*, 7(11): pp. 1218-1228.

Made in the USA
Las Vegas, NV
10 July 2024